AS/A-LEVEL YEAR 1

STUDENT GUIDE

OCR

Biology A

Module 2

Foundations in biology

Richard Fosbery

PHILIP ALLAN FOR
HODDER
EDUCATION
AN HACHETTE UK COMPANY

Philip Allan, an imprint of Hodder Education, an Hachette UK company, Blenheim Court, George Street, Banbury, Oxfordshire OX16 5BH

Orders

Bookpoint Ltd, 130 Milton Park, Abingdon, Oxfordshire OX14 4SB

tel: 01235 827827

fax: 01235 400401

e-mail: education@bookpoint.co.uk

Lines are open 9.00 a.m.–5.00 p.m., Monday to Saturday, with a 24-hour message answering service. You can also order through the Hodder Education website: www.hoddereducation.co.uk

© Richard Fosbery 2016

ISBN 978-1-4718-4390-7

First printed 2016

Impression number 5 4 3 2 1

Year 2020 2019 2018 2017 2016

This guide has been written specifically to support students preparing for the OCR AS and A-level Biology A examinations. The content has been neither approved nor endorsed by OCR and remains the sole responsibility of the author.

Cover photo: Argonautis/Fotolia; p. 8, Richard Fosbery; p. 10 top, Biology Pics/Science Photo Library; p. 10 bottom, Dr Jeremy Burgess/Science Photo Library; p. 83, Mike Samworth

Typeset by Integra Software Services Pvt. Ltd, Pondicherry, India

Printed in Italy

Hachette UK's policy is to use papers that are natural, renewable and recyclable products and made from wood grown in sustainable forests. The logging and manufacturing processes are expected to conform to the environmental regulations of the country of origin.

Contents

◼ Getting the most from this book

Exam tips

Advice on key points in the text to help you learn and recall content, avoid pitfalls, and polish your exam technique in order to boost your grade.

Knowledge check

Rapid-fire questions throughout the Content Guidance section to check your understanding.

Knowledge check answers

1 Turn to the back of the book for the Knowledge check answers.

Summaries

■ Each core topic is rounded off by a bullet-list summary for quick-check reference of what you need to know.

Exam-style questions

Commentary on the questions

Tips on what you need to do to gain full marks, indicated by the icon **e**

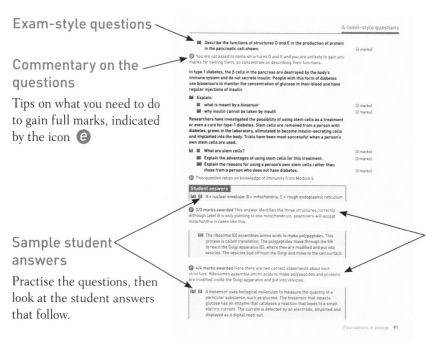

Commentary on sample student answers

Find out how many marks each answer would be awarded in the exam and then read the comments (preceded by the icon **e**) following each student answer showing exactly how and where marks are gained or lost.

Sample student answers

Practise the questions, then look at the student answers that follow.

■ About this book

This guide is the first in a series of four covering the OCR AS and A-level Biology A specifications. It covers Module 2: Foundations in biology, and is divided into two sections:

■ The **Content Guidance** provides key facts and key concepts, and links with other parts of the AS and A-level course. The links should help to show you how information in this module is useful preparation for other modules.

■ The **Questions & Answers** section contains two sets of questions, giving examples of the types of question to be set in the AS and A-level papers. There are some multiple-choice questions and some structured questions. The AS-style questions are followed by answers written by two students. These are accompanied with comments on the answers. The A-level-style questions are followed by model answers without comments.

This guide is not just a revision aid. It is a guide to the whole module and you can use it throughout the 2 years of your course if you decide to take the full A-level.

You will gain a much better understanding of the topics in Modules 3, 4, 5 and 6 if you read around the subject. I have suggested some online searches that you can do for extra information. Information you find online can help you especially with topics that are best understood by watching animations of processes taking place. As you read this guide remember to add information to your class notes.

The Content Guidance will help you to:

■ organise your notes and to check that you have highlighted the important points (key facts) — little 'chunks' of knowledge that you can remember

■ understand how these 'chunks' fit into the wider picture. This will help to support:
   – Modules 3 and 4, which are covered in the second student guide in this series
   – Modules 5 and 6, if you decide to take the full A-level course, which are covered in the third and fourth student guides in this series

■ check that you understand the links to the practical work, since you must expect questions on practical work in your examination papers. Module 1 lists the details of the practical skills you need to use in the papers.

■ understand and practise some of the maths skills that will be tested in the examination papers — look out for this icon for examples: ⊞

The Questions & Answers section will help you to:

■ understand which examination papers you will take

■ check the way questions are asked in the AS and A-level papers

■ understand what is meant by terms like 'explain' and 'describe'

■ interpret the question material — especially any data you are given

■ write concisely and answer the questions that the examiners set

Content Guidance

■ Cell structure

Microscopy

Key concepts you must understand

Size matters

It is difficult to imagine the range of sizes that biologists deal with. A blue whale can be as long as 30 m. The largest viruses are about 0.0004 mm. Many plant and animal cells are between 0.02 mm and 0.04 mm.

We use microscopes in biology because much of what we want to see is so small. Many cells, for example, are about 0.02 mm across. At best, our eyes can only make out objects that are about 0.1 mm in diameter, so using our eyes alone we would never see structures inside cells. The **light microscope (LM)** uses a beam of light that is focused by means of glass lenses. The **electron microscope (EM)** uses a beam of electrons focused by magnetic lenses.

Units

The units to use for measuring microscopic structures are the micrometre (μm) and the nanometre (nm). Remember:

- to convert millimetres to micrometres, multiply by 1000
- to convert micrometres to nanometres, multiply by 1000

Also remember:

- 1 μm (micrometre) = 0.001 mm; 1000 μm = 1 mm
- 1 nm (nanometre) = 0.001 μm; 1000 nm = 1 μm

Exam tip

Always measure in millimetres and not in centimetres. Make conversions between millimetres and micrometres by multiplying or dividing by 1000.

Key facts you must know

Resolution is the ability to see detail. The light microscope has a resolution of 0.0002 mm (2.0×10^{-4} mm in standard form) or 0.2 μm. This means that two points this distance apart are viewed as separate objects. Visible light has a wavelength of between 400 nm and 700 nm. Objects about half the size of the wavelength interrupt the rays of light and are resolved in the LM. However, anything smaller than 0.0002 mm is not visible because it is too small to interrupt the light. No matter how much a photograph taken through the LM is enlarged, small cellular structures are never visible.

Resolution The minimum size of an object that can be seen.

 1 mm expressed in standard form is 1.0×10^{-3} m and 1.0×10^{3} μm.
1 μm expressed in standard form is 1.0×10^{-3} mm and 1.0×10^{3} nm.

Magnification is the ratio between the actual size of an object and the size of an image, such as a photograph or a drawing.

Examiners may ask you to calculate magnifications or actual sizes. You should use these formulae:

$$\text{magnification} = \frac{\text{size of image}}{\text{actual size}}$$

$$\text{actual size} = \frac{\text{size of image}}{\text{magnification}}$$

With the LM, some structures, such as mitochondria, are just visible.

Exam tip

Use this triangle to help you remember how to calculate magnifications and actual sizes; for example:

$$\text{magnification} = \frac{\text{image size}}{\text{actual size}}$$

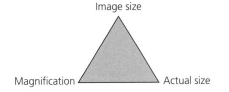

Image size

Magnification — Actual size

Electron microscopes

The wavelength of an electron beam is considered to be about 1 nm, so objects half this size are visible. As the resolution is so good, the magnification can be very high (×250 000 or more).

In the EM, magnets focus beams of electrons and an image is formed when the electrons strike a fluorescent screen or is captured by a sensor and displayed on a computer screen. The **transmission electron microscope (TEM)** is used to view thin sections of tissues. The **scanning electron microscope (SEM)** is used to view surfaces of three-dimensional objects, such as the bodies of insects and the surfaces of cells. In Figure 4 (p. 10) you can see the interior of a plant cell in 3D.

Inside electron microscopes is a vacuum, which allows electrons to travel towards the specimen. It means, however, that living cells cannot be observed, since these would explode. In the light microscope it is possible to watch living processes, such as cell division.

Table 1 The main characteristics of light and electron microscopes

Characteristic	Light microscope	Transmission electron microscope	Scanning electron microscope
Wavelength	400–700 nm	Varies depending on the voltage used in the microscope	
Effective maximum resolution	200 nm	0.2 nm	3.0 nm
Useful magnification	Up to ×1000 (at best ×1500)	Up to ×250 000 (or more) in TEM	Up to ×100 000

Magnification The ratio between the actual size of an object and an image of that object.

Knowledge check 1

The actual width of a plant cell is 40 µm. In an EM the width is measured as 75 mm. Calculate the magnification of the plant cell in the EM.

Knowledge check 2

The magnification of an animal cell in an EM is ×2000. The length of the cell in an EM is measured as 14 mm. Calculate the actual length of the animal cell.

Preparation of material for light microscopy

Most biological material is colourless or transparent and is composed of elements with low atomic mass. This means that visible light travels through tissues in the light microscope without being absorbed or reflected, so there is very little, if any, contrast. This problem is solved by adding stains, such as iodine, methylene blue and toluidine blue, all of which you may use during your course.

When iodine is added to some cells from a potato or a banana the nucleus, cell wall and cytoplasm stain yellow, but any starch grains inside the cells stain black. This is **differential staining** because the stain used gives two different colours. Often specimens for the LM are stained with several different stains, as in Figure 1.

Exam tip

Differential staining allows you to identify the different components of a cell in the light microscope.

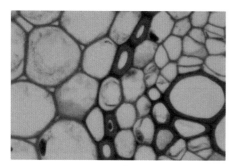

Figure 1 Differential staining of plant tissue — the cellulose cell walls are blue; cell walls strengthened with lignin are stained red (×250)

Staining is only part of the process of preparing slides for the LM. Small pieces of thin material are required as light rays do not penetrate very far through tissues. Some cells can be scraped or pulled away from the plant or animal tissue. When this is not possible, the fresh material can be cut into thin slices with a razor blade. These cells or thin sections are put on a slide, stained and covered with a cover slip.

If the material is delicate, it can be embedded in wax and then cut with a microtome, which is a specialised piece of apparatus that cuts sections that are all the same thickness. This can be used to make a series of sections across a structure such as a root or a blood vessel. The sections are then stained and mounted on slides. Apart from the temporary slides that you prepare yourself, the microscope slides you will see during your course will have been prepared like this.

Knowledge check 3

A student prepares a slide of plant tissue and stains it with iodine. Some areas stain blue-black and others stain yellow. Explain the different colours.

Exam tip

You should find and study many photographs taken with microscopes. This will help you to identify different types of cell and tissue.

Uses of microscopes and types of images

The LM is used for observing cells and tissues and for viewing structures that do not need to be seen in great detail. Living tissues can be seen, which is useful for viewing cells and tissues in action. Examples are:

- cytoplasmic streaming — movement of cell structures
- the coordinated movement of cilia, which move some small organisms
- the movement of chromosomes during nuclear division
- the separation of animal cells during cell division

Measurements of cells and cell structures are made with an eyepiece graticule. This is a graduated scale printed onto a piece of plastic that is inserted into the eyepiece. Before the scale is used it must be calibrated, because the actual width of the divisions on the scale depends on the magnification. As you increase the magnification, each division represents a smaller length.

A stage micrometer has a scale printed on it, usually with divisions that are 0.1 mm apart and 0.01 mm apart. The stage micrometer is put on the stage of the microscope and the eyepiece graticule is aligned with it, as shown in Figure 2. Count and record the number of lines on the graticule that correspond with 0.1 mm (or 0.01 mm) on the stage micrometer. When measuring the width of a cell with the LM, position the graticule over the object and count the number of graticule divisions.

Calculate the actual length using this formula:

$$\frac{\text{length of structure in graticule divisions}}{\text{number of graticule divisions equal to } 0.1\,\text{mm}}$$

Exam tip

There is a question on taking measurements with the LM on p. 69.

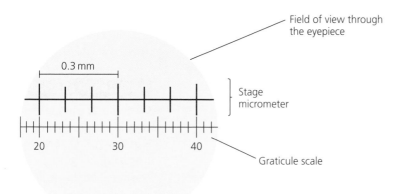

Figure 2 Calibrating an eyepiece graticule. The eyepiece is aligned with the stage micrometer to see how many divisions on the graticule represent 0.1 mm. 10 divisions on the graticule represent 0.3 mm. Each small graticule division = 0.03 mm or 30 μm.

Microscopes not only use visible light, but also ultraviolet light and lasers. The **laser scanning confocal microscope** allows images to be taken without light from areas that are out of focus. Computer technology is key to handling the images generated and presenting them to best effect. The cellular material is treated with different fluorescent stains so that different cellular components can be identified.

A **laser scanning confocal microscope** is a type of microscope that gives clear images without areas that are out of focus.

Exam tip

The electron micrographs that are used in examination papers are usually printed in black and white. Search online for black-and-white EMs of cells and organelles as part of your course and your revision.

Stains used in electron microscopy are salts of heavy metals, such as lead and uranium. These stains combine with proteins, for example in membranes, and absorb or scatter electrons as they pass through the specimen. This makes these areas show dark on images taken using the EM.

The stains used in electron microscopy do not give coloured images. Instead they are in black and white, but many electron micrographs are colour enhanced using computer software, as is the case with the TEM of the plant cell in Figure 3.

Exam tip

Search online for images taken with laser confocal microscopes and also see those prepared for fluorescence light microscopy. These are particularly good for viewing the cytoskeleton (Table 2 on pp. 12–13).

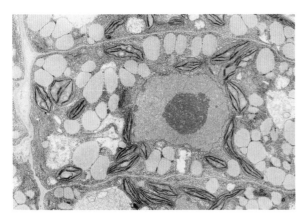

Figure 3 An electron micrograph of a plant cell, taken with a transmission electron microscope (×1600)

Exam tip

The cell walls are clearly visible in both EMs. In Figure 3 you can also see some cellular components: the nucleus, nuclear envelope, nucleolus, many chloroplasts and small vacuoles. In Figure 4 you can only see amyloplasts (organelles for the storage of starch).

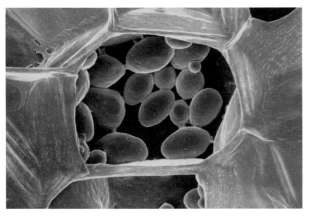

Figure 4 This electron micrograph shows a 3D image of a cell from a potato tuber, taken with a scanning electron microscope (×300)

Cell structure and function

Key facts you must know

Cell ultrastructure

The fine structure of cells that is visible with an electron microscope is often called **cell ultrastructure**. Many of the structures (**organelles**) within plant and animal cells, such as chloroplasts and mitochondria, are made of membranes. Within the cytoplasm of plant and animal cells are fibrous structures made of protein that make up the **cytoskeleton**.

Knowledge check 4

Calculate the actual widths of the cells in Figures 3 and 4. Show the details of your working and express your answer to the nearest micrometre.

Exam tip

If you calculate an actual size, check that it looks right. Examples of objects you may be asked about include: cells, 10–100 µm; chloroplasts, 3–10 µm; mitochondria, 1–3 µm; bacteria, 0.5–30 µm; membranes, 7–10 nm. If your answers are very different from these values, then you must have made a mistake. You may have to give an answer to the nearest whole number, which means rounding up or down.

Animal and plant cells are **eukaryotic** because they have a nucleus and organelles (Figures 5 and 6).

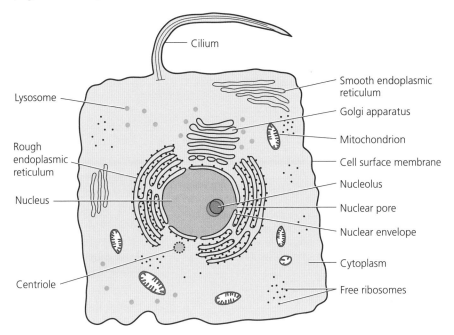

Figure 5 A generalised animal cell viewed with the transmission electron microscope

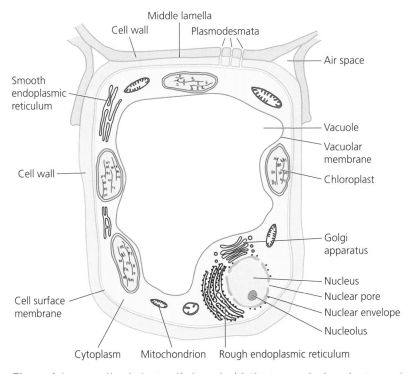

Figure 6 A generalised plant cell viewed with the transmission electron microscope

Functions of cellular components

Table 2 summarises the features and functions of the main cellular components in cells.

Table 2 Cellular components

Cellular components	Features	Function(s)
Rough endoplasmic reticulum (RER)	Flat sacs of membrane enclosing fluid-filled space; outer surface is covered in ribosomes	Ribosomes carry out protein synthesis; RER modifies proteins and transports them to Golgi apparatus
Smooth endoplasmic reticulum (SER)	Like RER but with no ribosomes on outer surface	Makes triglycerides (fats), phospholipids, cholesterol
Golgi apparatus	Pile of flat sacs with vesicles forming around the edge	Modifies proteins by adding carbohydrates, for example; packages proteins into vesicles for secretion; makes lysosomes
Mitochondria (singular: mitochondrion)	Formed of two membranes surrounding a fluid-filled matrix; inner membrane is highly folded to give large surface area for enzymes of respiration	Site of aerobic respiration; mitochondria have DNA and ribosomes and can make some of the proteins they use
Ribosomes	Attached to RER or free in cytoplasm — made of protein and RNA	Assemble amino acids to make proteins
Lysosomes	Single membrane surrounds fluid filled with enzymes	Contain enzymes for destroying worn-out parts of cell and for digesting food particles
Chloroplasts (plants and some protoctists only)	Many internal membranes giving a large surface area for chlorophyll, other pigments and enzymes of photosynthesis	Site of all the reactions of photosynthesis
Plasma (cell surface) membrane	Several — see pp. 49–51 for details	Controls entry and exit of materials; retains cell contents
Nucleus	Clearly visible in LM and EM when stained	Contains genetic information as DNA in chromosomes
Nuclear envelope	Structure like that of ER with ribosomes on outer surface; pores to allow substances to pass between cytoplasm and nucleus	Separates nucleus from cytoplasm
Nucleolus	Darkly staining area in nucleus	Produces ribosomes
Cytoskeleton microfilaments	Made of actin — a type of protein	Help to provide mechanical support for cells
Cytoskeleton microtubules	Made of tubulin — a type of protein that is formed into hollow tubes	Provide pathways within cells to enable vesicles and organelles to move about within the cytoplasm; form cilia, flagella and centrioles; form the spindle that moves chromosomes during anaphase of nuclear division →

Exam tip

Never write that mitochondria 'produce' or 'create' energy. Remember the first law of thermodynamics and never write about 'production of energy'.

Exam tip

You can learn the names of the organelles and their functions by copying this table and making drawings of the organelles. Divide them into organelles made of fibres and those made of membranes.

Cellular components	Features	Function(s)
Cilia (singular: cilium)	Extend from cell surface; made of microtubules in a '9+2' arrangement in the shaft (9 peripheral microtubules and 2 central ones), no central microtubules in the base; extend from cell surface; surrounded by plasma membrane	Found in large groups; move fluid or mucus past cells (e.g. in the trachea); move eggs along the fallopian tubes
Flagella (singular: flagellum)	As for cilia	Found singly to move individual cells, e.g. sperm in animals and in some plants such as ferns and mosses
Centrioles	Made of microtubules in same arrangement as in base of a cilium; not found in flowering plants	Form part of the centrosome in animal cells that organises microtubules to form the spindle during nuclear division (mitosis and meiosis)
Cell wall (plants and fungi; some protoctists)	Plant cell walls are made of cellulose; fungal cell walls are made of chitin	Withstands pressure of contents of cells

Exam tip

Look at images of cells to help you interpret photomicrographs, such as Figure 1, and electron micrographs, such as Figures 3 and 4.

Some of the organelles in Table 2 are involved in the production of protein. The diagram in Question 6 of the A-level-style paper on p. 90 shows you organelles, such as the nucleus, RER, the Golgi apparatus and secretory vesicles, working together to make and secrete a protein.

Key concepts you must understand

The cells depicted in Figures 5 and 6 are 'generalised' cells. They do not exist! They are drawn to show all the structures in plant and animal cells.

You should look carefully at photographs taken through the light microscope (these are known as photomicrographs or PMs) to see the differences between plant and animal cells. Sometimes you will be expected to identify organelles from electron micrographs (known as EMs) or from drawings made from electron micrographs. You should become proficient at recognising the organelles and using this information to explain how the structure of a cell, such as a sperm cell or a guard cell, is related to its function.

Links

Aspects of cell structure and function occur throughout the course. In Module 4 you will study the action of phagocytes and lymphocytes in defence against disease-causing organisms. You should look at PMs and EMs of these cells and relate their structure to their functions.

Prokaryotes

Key facts you must know

Prokaryotic cells (Figure 7) do not have a nucleus and there are no organelles made of membranes. Most prokaryotic cells are smaller than eukaryotic cells.

Links

Every time you come across cells of different types, check to see if they are eukaryotic or prokaryotic. Many of the parasitic organisms that cause diseases in plants and animals are prokaryotic, but there are many eukaryotic parasites, such as the one that causes malaria (infectious diseases are in Module 4).

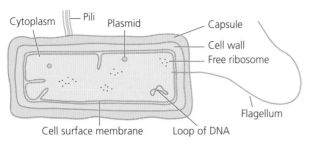

Figure 7 A generalised prokaryotic cell

Knowledge check 5

Make a table to show the differences between prokaryotic and eukaryotic cells.

Knowledge check 6

Make a table to show the differences between animal and plant cells.

Table 3 A summary of the differences and similarities between eukaryotic and prokaryotic cells

Structures shared with eukaryotic cells	Cytoplasm; ribosomes; plasma (cell surface) membrane. Ribosomes in prokaryotes are smaller than those in the cytoplasm of eukaryotes. They are 20 nm in diameter as opposed to eukaryotic ribosomes, which are about 30 nm.
Structures from eukaryotic cells never found in prokaryotic cells	Nucleus; nucleolus; nuclear envelope; mitochondria; Golgi apparatus; chloroplasts; flagella; cilia; vacuoles; linear DNA in chromosomes within nuclear envelopes; cell walls of cellulose
Structures only found in prokaryotic cells	Ring of DNA (sometimes called bacterial chromosome); cell walls of murein (a type of polysaccharide)
Structures found in some prokaryotic cells	Small rings of DNA known as plasmids; pili (small projections from the surface for exchanging DNA with other prokaryotes); slimy outer capsule for protection; flagellum (not built of microtubules)

Summary

- Light microscopes (LM) and electron microscopes (EM) are used to study cells, tissues and cell structures. Light microscopy allows us to see living cells, while transmission electron microscopy provides much greater detail of cells. Scanning electron microscopes (SEM) give details of the surfaces of structures.
- Stains are used in the LM because many structures are transparent; they are used in the EM to absorb or scatter electrons as compounds in biological specimens have atoms with low atomic mass and do not scatter electrons.
- The resolution of the LM is 200 nm and of the EM is about 0.2 nm; the maximum useful magnification for the LM is ×1000 and for the EM is ×250 000.
- Resolution is the ability to see detail; magnification is the ratio between image size and actual size.
- Organelles are sub-cellular components that perform specific functions for a cell. Some are made of membranes; those that form the cytoskeleton are protein fibres.
- Protein production and secretion is a function of many cells. The nucleus, ribosomes, RER, Golgi apparatus, secretory vesicles and plasma membrane function together to produce and export proteins.
- The cytoskeleton gives mechanical support to cells (as does the skeleton to the human body), moves organelles around the cell and brings about the movement of whole cells.
- Eukaryotic cells have a nucleus; prokaryotic cells do not. Eukaryotic cells have different types of membranous and fibrous sub-cellular components; prokaryotes have very few of these.
- There are many similarities between the structures of plant and animal cells; for example, both have a plasma (cell surface) membrane, mitochondria, ER, Golgi apparatus. Plant cells have cell walls, chloroplasts and large central vacuoles; animal cells do not.

■ Biological molecules

Water

Key concepts you must understand

It may seem odd to start a section on biological molecules with water, but in many ways it is the most important. Without it would there be life on Earth? We would expect water to be a gas at the temperatures on Earth. A heavier molecule with a similar formula, hydrogen sulfide (H_2S), is a gas. But most water is a liquid rather than a gas, because of hydrogen bonding.

Key facts you must know

Water molecules are dipolar (two poles). The electrons that form the covalent bond between hydrogen and oxygen tend to remain closer to the oxygen atom, giving it a slight negative charge (indicated by $\delta-$). The hydrogen atoms have a slight positive charge ($\delta+$), which means they are attracted to oxygen atoms on adjacent water molecules. This weak attraction between hydrogen and oxygen is called a hydrogen bond. Hydrogen bonding between water molecules is shown in Figure 8.

Roles of water in organisms

Reactant

Water takes part in chemical reactions; for example, it is a reactant in hydrolysis reactions in which covalent bonds between monomers in large molecules are broken. See pp. 18, 21 and 24 for examples. The digestion of complex molecules of food in animals relies on hydrolysis.

Solvent action

Ions (e.g. sodium and chloride ions) and polar molecules (e.g. glucose and amino acids) are charged. They are attracted to water molecules because of the weak positive and negative poles and are therefore dispersed easily in water to form solutions.

Water is a good solvent for ions and many biological molecules. Because of this water is a transport medium in the blood of animals and in the xylem and phloem of plants (see the transport section in the second student guide of this series).

Cohesion

Hydrogen bonds cause water molecules to 'stick together'. This makes it possible for them to travel up xylem vessels in plants in the transpiration stream (see the section on plant transport in the second student guide of this series).

High latent heat of vaporisation

It takes energy to break hydrogen bonds between water molecules and convert liquid water to water vapour. When water evaporates from plants and animals, it cools them down. This role as a coolant is useful for organisms that live in hot places.

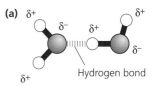

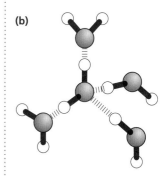

Figure 8 (a) Two water molecules with a hydrogen bond between them (b) A cluster of water molecules held together by hydrogen bonds

Exam tip

Ions and polar molecules are important in biology. Make sure that you understand why they are charged and how this makes them soluble in water.

High latent heat of fusion

Much thermal energy is needed to change ice to liquid water; much is transferred from water when it forms ice. Water in cells tends to stay as a liquid, so cell membranes are not damaged by ice crystals.

Specific heat capacity

Water absorbs a significant amount of energy before it changes state, so the temperature does not change quickly. This provides the relatively stable habitats provided by bodies of water — sea water and freshwater — for many species of aquatic eukaryotes and prokaryotes.

Links

Hydrogen bonds are important in stabilising proteins (p. 25), maintaining the structure of DNA and tRNA and forming strong molecules, such as cellulose. Before DNA can be replicated, hydrogen bonds between the two polynucleotide chains making up the double helix must be broken (p. 33). The transport of water in xylem tissue in plants relies on hydrogen bonding.

Knowledge check 7

List three roles of water in mammals.

The chemistry of biological molecules

The biological molecules in the sections that follow are much more complex than water. The major chemical elements that are in biological molecules are carbon, hydrogen, oxygen, nitrogen, sulfur and phosphorus. Table 4 shows the elements in each of the four groups of biological molecule.

Table 4 The chemical elements in the four groups of biological molecules

Element	Carbohydrates	Lipids	Proteins	Nucleic acids
Carbon	✓	✓	✓	✓
Hydrogen	✓	✓	✓	✓
Oxygen	✓	✓	✓	✓
Nitrogen	✗	✗	✓	✓
Sulfur	✗	✗	✓	✗
Phosphorus	✗	✗	✗	✓

Macromolecules

The four groups in Table 4 include very large molecules or **macromolecules**. These are made from smaller molecules that are bonded together. As you will see over the next few pages, large carbohydrates (e.g. starch and cellulose), proteins and nucleic acids are composed of repeating molecules that are bonded together. Macromolecules like this are **polymers**, which are made from many identical or similar **monomer** molecules. Lipids are unlike the others as they do not have repeating monomers.

Exam tip

Some biological polymers are made of repeating monomers, i.e. AAAAAA, as in the case of starch. Others are made of two or more monomers of similar structure, e.g. ABABAB… or ABCDBDADC… To see the difference between lipids and the other macromolecules contrast Figure 16 with Figure 14.

A **monomer** is a small molecule (e.g. glucose, amino acids) that can be bonded together to form a polymer.

A **polymer** is a large molecule formed of many monomers bonded together, e.g. protein formed from many amino acids.

A **macromolecule** is a large molecule that may or may not be a polymer.

Carbohydrates

Key concepts you must understand

Simple carbohydrates, such as glucose, can be joined together to form large molecules. Starch, cellulose and glycogen are polymers made of many molecules of glucose joined together. The 3-D structures of these large carbohydrates give them special properties and functions in cells.

Long-chain molecules, such as starch and glycogen, are good for storage since they are compact. Amylopectin and glycogen are branched with numerous endings where glucose molecules can be added when they are available and where they can be removed when glucose is required for respiration.

Key facts you must know

Carbohydrates:
- contain hydrogen and oxygen in a ratio of 2:1 — for example glucose, $C_6H_{12}O_6$
- have the general formula $C_x(H_2O)_y$ in which x can be 3, 4, 5, 6 or 7
- include **monosaccharides**, sometimes known as simple sugars, for example glucose and ribose; **disaccharides**, also known as complex sugars, for example sucrose; and **polysaccharides**, for example starch (amylose and amylopectin), glycogen and cellulose

Ribose (Figure 9) is a pentose sugar with five carbon atoms. It is one of the components of the monomers that are polymerised to form ribonucleic acid (p. 30).

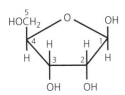

Figure 9 A molecule of ribose, a pentose with five carbon atoms; deoxyribose has the same structure except at C2 it has –H not –OH

Glucose is a hexose sugar with six carbon atoms. The –OH groups that surround the molecule are polar so they form hydrogen bonds with water. This means that it dissolves in water fairly readily. As it is relatively small it passes through carrier proteins in plasma (cell surface) membranes easily. It is also easily metabolised by converting to larger compounds for storage (see p. 19) and for respiration to release energy. Figure 10 shows the two forms of glucose: alpha (α) and beta (β).

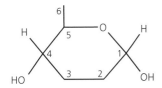

α-Glucose (note that the –H is above the –OH on carbon atom 1)

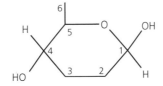

β-Glucose (note that the –OH is above the –H on carbon atom 1)

Figure 10 Two forms of glucose: α-glucose and β-glucose

The difference between the two is very small, yet polymers of these two forms of glucose show important differences.

Exam tip

There are two other hexose sugars that you should know: galactose and fructose. The other pentose you should know is deoxyribose. There is more about these sugars later.

Knowledge check 8

State the differences between the structure of α-glucose and β-glucose.

Knowledge check 9

State two ways in which the structure of an α-glucose molecule differs from that of a ribose molecule.

Content Guidance

Maltose, sucrose and lactose are disaccharides as they are composed of two simple sugar units joined together by a glycosidic bond. Two α-glucose molecules are joined together to form maltose:

glucose + glucose → maltose

The glucose molecules are joined together by a **glycosidic bond** between carbon atom 1 on one glucose and carbon atom 4 on another. The bond is called a 1,4-glycosidic bond.

Figure 11 shows how glucose is joined with another simple sugar, fructose, to form sucrose:

glucose + fructose → sucrose

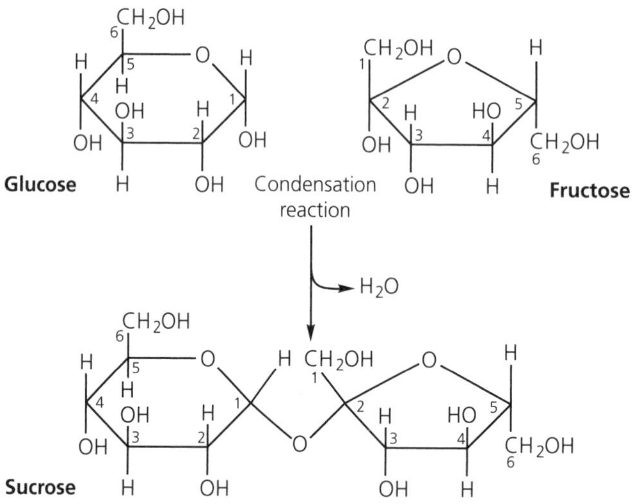

Figure 11 Forming a 1,2-glycosidic bond between glucose and fructose to form sucrose

Exam tip

Maltose is always made as a product of breakdown reactions so the reaction shown here is purely hypothetical. That does not stop examiners asking questions about it.

This type of reaction is a condensation or dehydration reaction, because a molecule of water is formed. Figure 12 shows the hydrolysis of sucrose when it reacts with water. It is what happens when sucrose is boiled with hydrochloric acid (see p. 28).

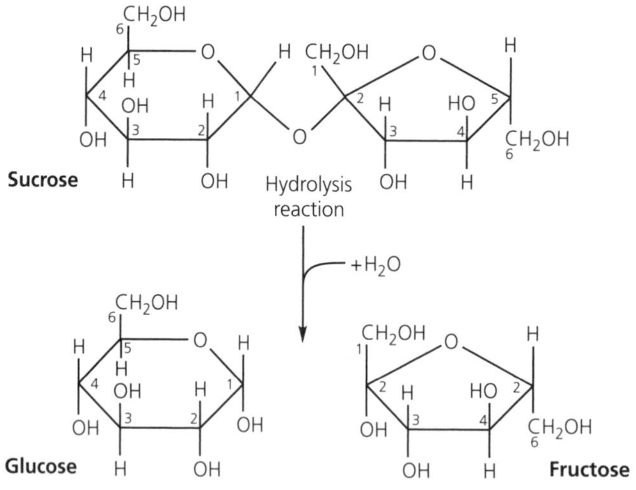

Figure 12 Breaking the 1,2 glycosidic bond in sucrose to make glucose and fructose

The acid acts as a catalyst, speeding up the addition of water to break the glycosidic bond.

18   OCR Biology A

Lactose is another sugar formed from glucose and galactose. It is only known in mammalian milk and no one knows quite why this is. Glucose would be just as good, if not better. Maltose, sucrose and lactose are disaccharides as they are composed of two molecules of simple sugars.

Starch (amylose and amylopectin), glycogen and cellulose are polymers of glucose. They are made of many glucose molecules joined by glycosidic bonds. Amylose is formed when α-glucose monomers, joined by 1,4-glycosidic bonds, make a long chain with a compact helix structure. Amylopectin and glycogen are also polymers of α-glucose, but some of the glucose molecules are attached by 1,6-glycosidic bonds to form branching points (Figure 13).

Amylose, amylopectin and glycogen are storage molecules. They are good for this as they are insoluble and can store many thousands of molecules of glucose. They are compact, so do not take up much space within cells. Glycogen and amylopectin are highly branched, which gives each molecule many 'ends' where glucose can be added or removed.

Exam tip

In the diagram on p. 93 (Question 7 of the A-level-style questions) you can see the structure of lactose (a disaccharide) and the two monosaccharides that compose it.

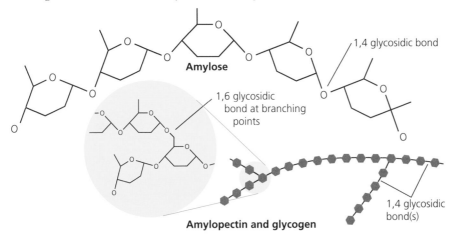

Figure 13 Branching points in amylopectin and glycogen are formed by 1,6-glycosidic bonds

Cellulose (Figure 14) is an unbranched polymer of thousands of β-glucose molecules. Because of the arrangement of the −OH and −H groups about C1, each monomer is arranged at 180° to those on either side. This arrangement does not form a helix, like amylose, but a straight chain. The molecules of cellulose are laid down in parallel to form microfibrils. Many hydrogen bonds form between the many −OH groups that project from all sides of the molecule. This gives the microfibrils great strength. Bundles of microfibrils are aligned in cell walls to resist the turgor pressure inside plant cells.

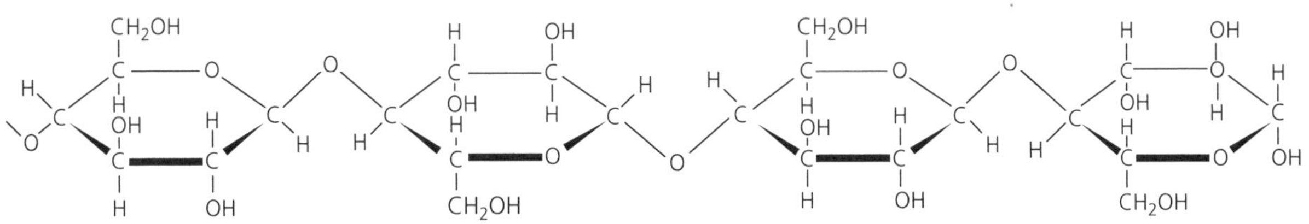

Figure 14 A molecule of cellulose. The last monomer in the chain is on the right-hand side. Notice that the β-glucose units are arranged at 180° to each other

Table 5 summarises some features of four polysaccharides.

Table 5 The main features of four polysaccharides

Polysaccharide	Monomer	Glycosidic bond(s)	Where found	Functions
Amylose	α-glucose	1,4 (unbranched helical molecule)	Amyloplasts (see Figure 4, p. 10) and starch grains in chloroplasts	Energy store in plants
Amylopectin	α-glucose	1,4; 1,6 (branched molecule)	As for amylose	Energy store in plants
Glycogen	α-glucose	1,4; 1,6 (branched molecule)	Granules in animal cells (e.g. liver and muscle)	Energy store in animals
Cellulose	β-glucose	1,4 (straight-chain molecule)	Cell walls of plants	Structural — making cell walls in plants

Links

When enzymes break down sucrose and starch, they catalyse the hydrolysis of glycosidic bonds. You will learn more about the production of sugars during photosynthesis in Module 5 (Student Guide 3 in this series). Triose (3C) sugar molecules produced inside chloroplasts are transferred to the rest of the cell and used in respiration or used to make sucrose, starch or cellulose. Sucrose can be exported to the rest of the plant in the phloem (see the plant transport section in Student Guide 2 in this series).

Lipids

Key concepts you must understand

Lipids are macromolecules. They are not polymers as they do not have repeating sub-units. Lipids are a large, diverse group of compounds. Examples include triglycerides (fats and oils), phospholipids, steroids and waxes. Lipids are not soluble in water; they are soluble in organic solvents, such as ethanol.

Key facts you must know

Like carbohydrates, lipids are composed of carbon, hydrogen and oxygen, but these elements are in different proportions. In lipids, there is far more hydrogen than oxygen. The major components of triglycerides (fats and oils) and phospholipids are fatty acids that have long hydrocarbon chains (Figure 15). The other main component is glycerol.

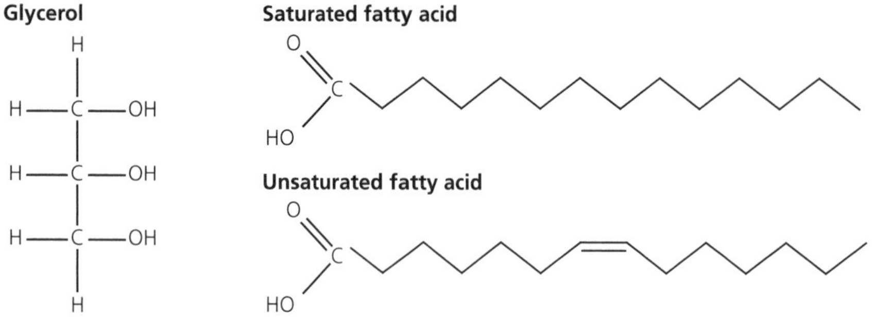

Figure 15 Saturated fatty acids have no double bonds between the carbon atoms, while unsaturated fatty acids have one or more double bonds

Knowledge check 10

State the products of the complete hydrolysis of maltose, sucrose, glycogen, amylopectin, amylose and cellulose.

Knowledge check 11

Describe the differences between the structures of amylose and cellulose. Explain how the structures of these two molecules are related to their functions in plant cells.

Exam tip

A double bond is a covalent bond in which four electrons are shared between atoms rather than the two shared in a single bond. A double bond is stronger than a single bond.

Triglycerides

The fatty acids in a triglyceride can be the same as, or different from, each other. The bond that forms between a fatty acid and glycerol is an ester bond (Figure 16). During the formation of an ester bond, water is eliminated — another example of a condensation reaction.

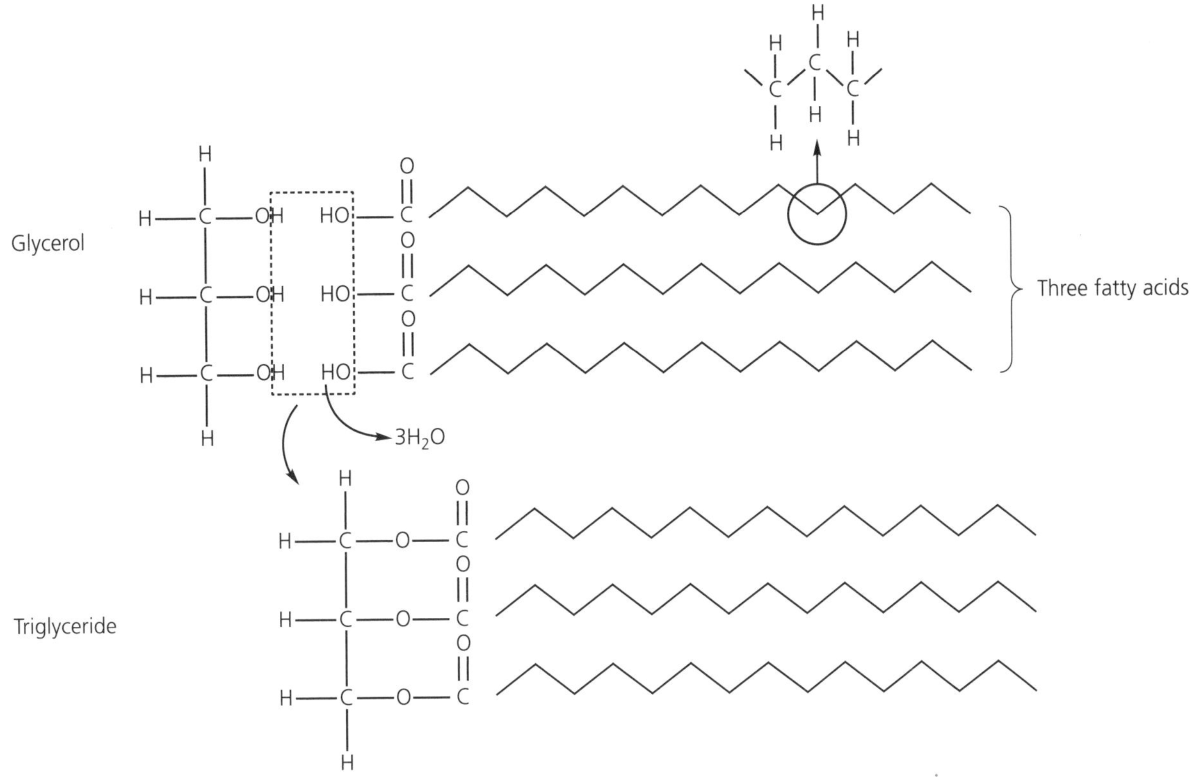

Figure 16 The formation and breakdown of a triglyceride molecule by the formation of ester bonds (esterification)

Triglycerides have no polar groups so are hydrophobic and not soluble in water. They are energy-rich and are therefore excellent for energy storage. When respired, the molecule is oxidised, releasing much energy and hydrogen atoms, which form water with oxygen. This is metabolic water, which is an important source of water for desert animals, such as gerbils and camels.

Fat is less dense than water, and so gives buoyancy to aquatic mammals, such as dolphins and whales. It is also a poor conductor of heat, so is an excellent thermal insulator. Its soft, cushioning effect makes it good for protecting organs, such as the kidneys.

Many seeds are rich in triglycerides as it is an energy-dense substance — the energy released per gram is the highest of all the biological molecules.

Knowledge check 12

Triglycerides and glycogen are used for energy storage in animals. State three ways in which the structure of a triglyceride differs from that of glycogen.

Phospholipids

All phospholipids have a phosphate group attached to glycerol as well as two fatty acids (Figure 17). In most phospholipids, the phosphate is often attached to nitrogen-containing, polar groups. One of these is choline. The phosphate group and any other groups attached to it make the 'head' of the molecule 'water loving' or **hydrophilic**. The two fatty acid chains do not 'like' water — they are **hydrophobic**.

Hydrophilic or 'water loving' substances can interact with water molecules and dissolve in water.

Hydrophobic or 'water hating' substances cannot interact with water molecules and are insoluble in water.

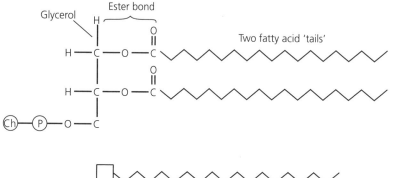

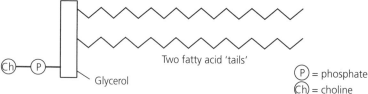

Figure 17 Two ways of showing a phospholipid molecule

Phospholipids form two layers (a bilayer) to make membranes. They are ideal for this, as the hydrophilic regions interact with water in the cytoplasm and in the fluid outside the cell, while the fatty acid chains form a hydrophobic core. This makes a barrier to the movement of water-soluble substances in and out of cells as well as organelles made of membranes, such as mitochondria, lysosomes, chloroplasts and endoplasmic reticulum.

Exam tip

The polar regions of phospholipids form hydrogen bonds with water in fluids on either side of a bilayer.

Exam tip

Look at the structure of a phospholipid molecule and then look at the diagram of a cell surface membrane in Figure 41. Phospholipid molecules are suitable for making a bilayer with a hydrophobic core that separates two areas dominated by water.

Cholesterol

Cholesterol (Figure 18) is a lipid-like molecule that is like phospholipids in that it has hydrophobic and hydrophilic regions and is an important component of membranes of eukaryotic cells. It is also used to make steroids, such as testosterone and progesterone.

The liver is responsible for metabolising cholesterol for the body. As it is needed everywhere in the body it has to be transported in the blood, but it is not water soluble, so cannot simply dissolve in the blood plasma.

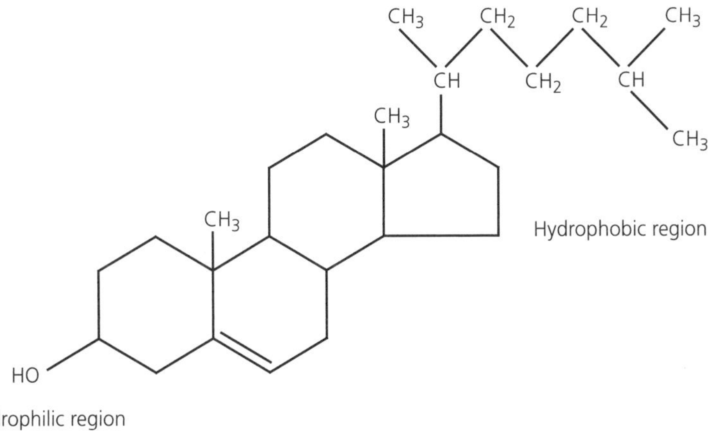

Figure 18 Cholesterol

Knowledge check 13

a Describe how an unsaturated fatty acid differs from a saturated fatty acid.
b State *three* distinctive features of cholesterol.

Knowledge check 14

Make a table to compare the structure and functions of triglycerides and phospholipids.

Amino acids and proteins

Key concepts you must understand

Proteins are unbranched macromolecules made of amino acids joined together by peptide bonds. Proteins are made from 20 different amino acids, which can be arranged in many different sequences. This gives a huge variety of different proteins having many different functions.

Single chains of amino acids are polypeptides. These may be folded into complex 3D shapes to form globular proteins, such as haemoglobin. Fibrous proteins, such as collagen, are made of polypeptides arranged in simpler shapes, such as helices.

Key facts you must know

Amino acids

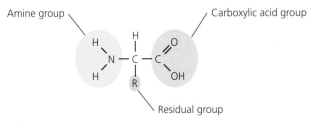

Figure 19 The generalised structure of an amino acid

Exam tip

Lipase is the enzyme that digests triglycerides. You can follow the digestion of fat by lipase with a pH meter or a pH indicator. This is possible because as fat is hydrolysed fatty acids are released, which lower the pH.

Exam tip

There are many more amino acids that are not used to make proteins. Ornithine and citrulline are two of these; they are part of the urea cycle (see Student Guide 3 in this series).

It is the R groups that make amino acids different from one another (Figure 19). Glycine (the simplest amino acid) has hydrogen (–H) as its R group. Alanine has a methyl group (–CH₃). Two amino acids are joined by a peptide bond that forms between the amino group of one amino acid and the carboxylic acid group of the other (Figure 20). This is another example of a condensation reaction.

Knowledge check 15

Explain how it is possible for organisms to make a huge variety of different protein molecules, but only a much smaller number of polysaccharides.

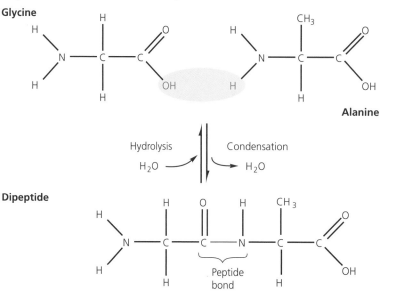

Figure 20 Formation and breakage of a peptide bond between glycine and alanine

Some proteins, such as the enzymes lipase and lysozyme, are made of one polypeptide; some are made of two or more. Insulin has two polypeptides joined by disulfide bonds; haemoglobin, the enzyme catalase and the antibodies known as IgG are made of four polypeptides. Levels of organisation in proteins refer to their structure (Figure 21 and Table 6).

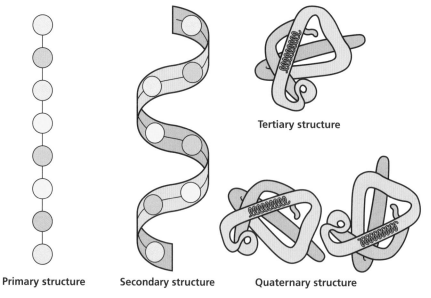

Figure 21 Levels of organisation in proteins

Table 6 Levels of organisation in proteins

Level of organisation	Structure
Primary structure	The sequence of amino acids in a polypeptide Amino acids are linked together by peptide bonds Also includes the position of disulfide bonds in polypeptides
Secondary structure	Polypeptide folded into an α-helix (a right-handed helix), or a β-pleated sheet Secondary structure is stabilised by hydrogen bonds between the –C=O and the –NH groups of peptide bonds (Figure 20)
Tertiary structure	Secondary structure folded to form complex 3D shape held together by a variety of bonds — see below
Quaternary structure	Two or more polypeptides arranged together

Exam tip

Search online for models of protein structure. You should find plenty of ribbon models that show secondary structure. Note that β-pleated sheets are depicted as broad arrows, as in Figure 23 (p. 26).

The tertiary structures of polypeptide chains are held in specific shapes by intramolecular bonds that exist within the molecule. Many of these bonds occur between the R groups that project from the central core of the molecule, formed by the carbon and nitrogen atoms (e.g. −CCNCCNCCNCCN−).

The disulfide bond is the strongest of these because it is a covalent bond. It is formed between R groups of the amino acid cysteine. The R group is −SH and when proteins are formed, these groups react to form the −S−S− bond shown in Figure 22. Disulfide bonds are common in proteins on the outside of plasma membranes (e.g. receptors) and those released into tissue fluid and blood (e.g. antibodies and insulin).

Ionic and hydrogen bonds break more easily, for example when a protein is heated above about 40°C or when it is exposed to a change in pH.

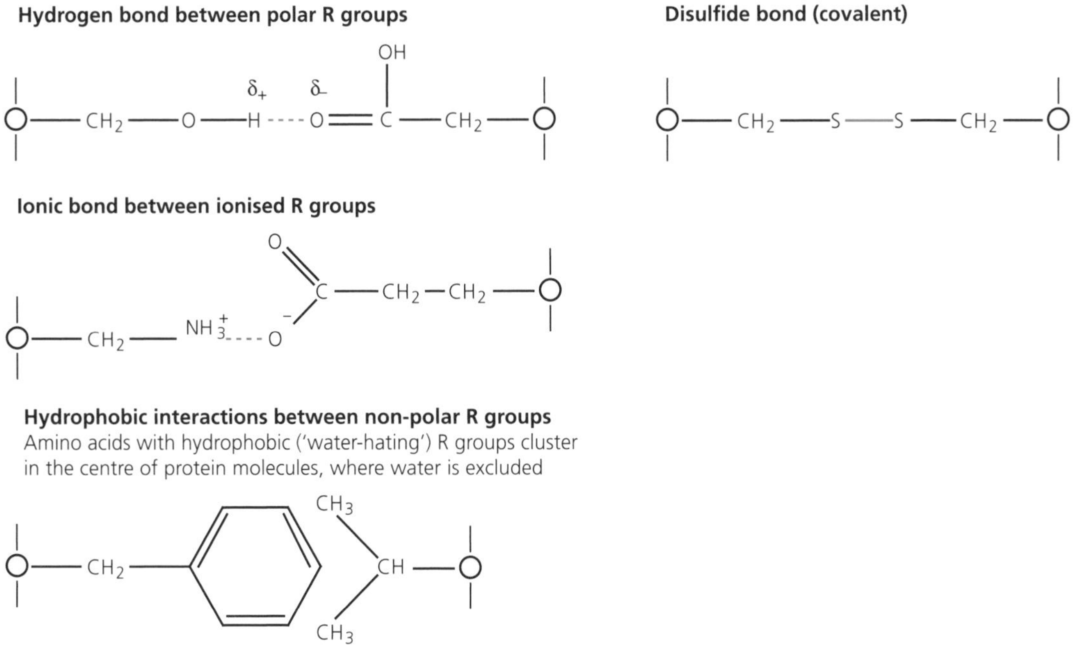

Hydrogen bond between polar R groups

Disulfide bond (covalent)

Ionic bond between ionised R groups

Hydrophobic interactions between non-polar R groups
Amino acids with hydrophobic ('water-hating') R groups cluster in the centre of protein molecules, where water is excluded

Figure 22 Intramolecular bonds that stabilise proteins

Globular proteins

Globular proteins have spherical shapes, with hydrophilic R groups projecting from the surface, so that they are water soluble. Often regions within the molecule are formed from amino acids with hydrophobic R groups, for example the 'pocket' where the haem group is attached to the polypeptides that compose haemoglobin.

Table 7 summarises the structure and function of three globular proteins: haemoglobin, lysozyme (Figure 23) and insulin.

Table 7 The structure and roles of three globular proteins found in mammals

Features	Haemoglobin	Lysozyme	Insulin
Number of polypeptides	Four: two α-globins and two β-globins	One	Two: A and B
Secondary structure	Each polypeptide has seven α-helices and no β-pleated sheets	The polypeptide has α-helices and β-pleated sheets	The A polypeptide has two α-helices and the B polypeptide has an α-helix and a β-pleated sheet
Prosthetic group	Each polypeptide has a haem group	None	None
Site of synthesis	Maturing red blood cells in bone marrow	For example, cells lining tear ducts; phagocytes	β cells in islets of Langerhans in the pancreas
Role	Transport of oxygen and carbon dioxide	Destroying cell walls of bacteria	Hormone: stimulates a decrease in blood glucose concentration
Site of action	Red blood cells	Many places in the body — for example, in phagocytes; in tears, sweat and saliva	On the surfaces of liver, muscle and fat cells throughout the body

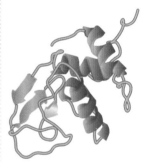

Figure 23 A ribbon model of a molecule of lysozyme

Haemoglobin is found in large quantities inside red blood cells. Each molecule of haemoglobin has four polypeptides: two α-polypeptides and two β-polypeptides. Each of the four polypeptides is attached to an iron-containing haem group. Oxygen attaches loosely to these haem groups to form oxyhaemoglobin. These haem groups are examples of **prosthetic groups**. A prosthetic group is any non-protein group that is permanently bound to a protein. Proteins that have prosthetic groups are known as **conjugated proteins**.

Fibrous proteins

Collagen, keratin and elastin are fibrous proteins. These proteins are not soluble in water.

- **Collagen** is an extracellular protein that provides toughness to skin, bone, cartilage, tendons, ligaments and muscles. The structure of each collagen molecule is three polypeptides wound tightly around each other in a triple helix. This helps to make collagen suitable for tendons, which need high tensile strength to resist pulling forces.

- **Keratin** also has helical molecules that coil around each other. It also has many disulfide bonds to make it a very tough substance that is found in hair and skin.
- **Elastin** is a protein that, unlike collagen and keratin, lengthens when stretched and then recoils once the pulling force stops. This property makes it suitable for the airways and alveoli in the lungs, which recoil when breathing out, and in arteries, which stretch and recoil as blood surges through them.

Exam tip

It is a good idea to look at computer models of the biological molecules, especially proteins. You can search for them online and display them. Or you could use your gaming skills at the website *Foldit*, which is dedicated to this. The site has links to images and information about proteins in your course at the Protein Data Bank.

Inorganic ions

An ion is a charged atom or a charged group of atoms. There are many ions that are involved with biological molecules in processes that occur inside and outside cells. Cations are positively charged ions; anions are negatively charged ions.

Table 8 shows the names and chemical symbols for several of the more important cations and anions, together with some of their roles.

Table 8 Important inorganic ions and some of their roles in organisms

Ions	Named ion	Chemical symbol	Roles
Cations	Calcium	Ca^{2+}	Strengthens bones and teeth; intracellular signalling; as calcium pectate holds plant cells together
	Sodium	Na^+	Nerve impulse transmission; important constituent of body fluids, for example blood, tissue fluid, sweat and tears
	Potassium	K^+	Opening of guard cells (in plants)
	Hydrogen	H^+ (protons)	Production of ATP by chemiosmosis in chloroplasts, mitochondria and prokaryotic cells
	Ammonium	NH_4^+	Making amino acids (in plants); converted into urea for excretion by mammals
Anions	Nitrate	NO_3^-	Absorbed by roots as a source of nitrogen for making amino acids and nucleotides (plants)
	Hydrogencarbonate	HCO_3^-	Transport of carbon dioxide (animals)
	Chloride	Cl^-	Maintaining neutrality in red blood cells (Bohr shift); cofactor of salivary amylase
	Phosphate	PO_4^{3-}	Component of nucleotides, including ATP, and nucleic acids
	Hydroxide	OH^-	Produced in immune responses to kill infected cells; these ions damage macromolecules such as DNA

Exam tip

You will come across some of these roles later in your course. Make a copy of this table and add other roles as you come across them.

Exam tip

Any charged compound with at least one C–H bond is an organic ion. Examples are amino acids and carboxylic acids, such as lactic acid. They are usually charged at the pH of cytoplasm and biological fluids, such as blood.

Genes determine the primary structure of polypeptides. The relationship between the sequence of bases in the nucleic acids DNA and RNA and the sequence of amino acids in a polypeptide is explained on p. 35.

The specific functions of an enzyme are determined by its tertiary structure. The tertiary structure is determined by the way in which the sequence of amino acids folds, and this is partly determined by the relative positions of amino acids with different types of R group. Some R groups are polar and hydrophilic; others are non-polar and hydrophobic.

Chemical tests for biological molecules

Table 9 summarises the tests for biological molecules that you should know.

Table 9 Chemical tests for some biological molecules

Test substance	Reagent	Details of test	Positive result	Negative result
Starch	Iodine solution	Add to solid or to a solution	Blue-black colour	Yellow colour
Reducing sugar	Benedict's solution	Add to a solution; boil (or put at >80°C)	Green, yellow, orange, red colour with precipitate	No change to blue colour; no precipitate
Non-reducing sugar	Hydrochloric acid; sodium hydroxide solution Benedict's solution	Sample 1 — as per reducing sugar Sample 2 — boil with HCl and then neutralise with NaOH; boil with Benedict's solution	Sample 1 – no change to blue colour (if no reducing sugars) Sample 2 — any colour change (as for reducing sugar)	Both samples — no colour change
Protein	Biuret solution	Add to a solution	Violet/lilac/purple	No change to blue colour
Fat	Ethanol	Dissolve test substance in ethanol, then pour into water	Cloudy emulsion	No emulsion

Biosensors are used to determine the concentration of biologically important substances. Diabetics use biosensors to measure the concentration of glucose in their blood. They put a drop of blood onto the pad of a test strip and place it into the biosensor. An enzyme in the pad catalyses a reaction that generates a small electric current. This is amplified and interpreted as a concentration, which is visible as a digital display.

Chromatography is a technique for separating, identifying and measuring substances extracted from biological material. The substances being analysed are extracted in a suitable solvent and drops of the solution are placed on chromatography paper or a layer of fine-grained material, such as silica gel, on a solid support — a process known as thin-layer chromatography.

Exam tip

Knowing the structure of haemoglobin will help you to understand the way in which it works to transport oxygen. This is an important topic in Module 3 (see Student Guide 2 in this series).

Exam tip

Chromatography of vitamins is described in Question 8 of the A-level-style questions on p. 96. There you will find out how to calculate retention factors (Rf).

Summary

- Water molecules are dipolar and are attracted to each other by hydrogen bonds. This makes water a liquid over a wide range of temperatures and gives water many special properties that contribute to its roles in living organisms.

- Biological molecules are made from six main elements: carbon, hydrogen, oxygen, nitrogen, sulfur and phosphorus. Macromolecules are large and composed of sub-unit molecules that are joined together. Polymers, such as proteins, large carbohydrates and nucleic acids, are molecules made from many such sub-unit molecules known as monomers. Lipids are also macromolecules, but not polymers. These sub-unit molecules are joined together by condensation reactions in which water is formed.

- Monosaccharides are carbohydrates with three to seven carbon atoms. Ribose is a pentose with five; glucose is a hexose with six. In α-glucose the –H is above the –OH on carbon 1. In β-glucose the –H is below the –OH on carbon 1. Maltose, sucrose and lactose are disaccharides formed by condensation from two monosaccharides. Starch and glycogen are polymers of α-glucose. Cellulose is a polymer of β-glucose.

- Glycosidic bonds form between monosaccharides: 1,4 bonds give unbranched chains in amylose and cellulose; 1,6 bonds give branching points in amylopectin and glycogen.

- Amylose is a helical shape and is used as an energy store. Cellulose molecules are straight and form hydrogen bonds with adjacent molecules, forming strong microfibrils in cell walls to withstand the turgor pressures of plant cells.

- A triglyceride is formed from glycerol and three fatty acids by condensation to form ester bonds. A phospholipid has one fatty acid replaced by phosphate, often linked to another water-soluble group, such as choline.

- Triglycerides are insoluble in water and are used for long-term storage of energy. Their high hydrogen content gives high energy release when respired. Phospholipids have hydrophilic and hydrophobic regions, resulting in the formation of bilayers in cell membranes; cholesterol molecules with similar regions help to stabilise phospholipid bilayers.

- Amino acids are the monomers of proteins. A peptide bond is formed by a condensation reaction between the amine group of one amino acid and the carboxylic acid group of another. The products of such condensation reactions are dipeptides and polypeptides. The sequence of amino acids in polypeptides is the primary structure.

- The secondary structure (α-helix and β-pleated sheet) is the result of folding of a polypeptide. Further folding into complex shapes is the tertiary structure. Proteins with more than one polypeptide have quaternary structure.

- Proteins are stabilised by intramolecular bonds — hydrogen bonds, ionic bonds, disulfide bonds — and by hydrophobic and hydrophilic interactions. Hydrogen bonds and ionic bonds break when proteins are heated to high temperatures or are in extremes of pH.

- Haemoglobin, enzymes (e.g. lysozyme) and insulin are examples of globular proteins that are spherical and water soluble. Collagen, keratin and elastin are fibrous proteins that are insoluble, and provide support to tissues.

- Water is used in hydrolysis reactions to break down polymers and lipids to their constituent sub-unit molecules.

- Ions have different structural and functional roles within organisms. Many are used to make components of biological molecules. For example, nitrate ions are used by plants to make amino acids.

- Chemical tests are used to identify biological molecules. Starch is identified using iodine solution; reducing and non-reducing sugars using Benedict's solution; proteins using biuret solution. Lipids are identified by dissolving in ethanol and adding the solution to water to form an emulsion. Reagent test strips are used to detect different molecules including glucose.

- Colorimeters and biosensors are used to obtain quantitative data about the concentrations of biological molecules.

- Chromatography is used to separate the components of mixtures and identify them. The retention factor (Rf) of each compound is the ratio between the movement of the substance and the movement of the solvent front.

■ Nucleotides and nucleic acids

DNA and RNA — the nucleic acids

Key concepts you must understand

Nucleic acids are essential to heredity in living organisms. Understanding the structure of nucleotides and nucleic acids allows an understanding of their roles in the storage and retrieval of genetic information and controlling cell metabolism.

DNA (deoxyribonucleic acid) is a large, very stable molecule found in chromosomes. It forms the **genes** that code for proteins. You inherited it from your parents and you will pass it on to your children. It is a long-term store of genetic information. RNA (ribonucleic acid) is a shorter molecule that cells use to retrieve information from DNA and express it in the form of proteins.

A gene is a length of nucleotides with a specific sequence of bases. A gene determines the sequence of amino acids in a polypeptide, coding for the order in which the amino acids are put together. In DNA there are four nitrogenous bases (A, T, C and G), which cells read in groups of three. These triplets of bases code for the 20 different types of amino acid used to make proteins.

> A **gene** is a length of DNA that has a base sequence that determines the sequence of amino acids in a polypeptide.

Key facts you must know

DNA and RNA are macromolecules made of repeating sub-units joined together by covalent bonds. Nucleotides are the sub-unit molecules of nucleic acids (Figure 24). Cells join these nucleotides together by forming phosphodiester bonds to make polynucleotides.

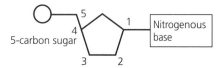

Figure 24 All nucleotides have this structure (the box represents one of five different nitrogenous bases, as in Table 10). The numbers indicate the positions of carbon atoms in the pentose (5C) sugar

The nucleotides that make up DNA contain the pentose deoxyribose; the nucleotides that make up RNA contain ribose. The group attached to C2 in ribose is $-OH$; in deoxyribose it is $-H$ (see Figure 9 on p. 17). Table 10 summarises the distribution of the two groups of bases: purines and pyrimidines. Purine bases have two rings of carbon and nitrogen atoms; pyrimidines have one.

Table 10 The nitrogenous bases in DNA and RNA

	Purines		Pyrimidines		
Number of rings	2		1		
Bases	Adenine	Guanine	Cytosine	Thymine	Uracil
DNA	✓	✓	✓	✓	✗
RNA	✓	✓	✓	✗	✓

In both DNA and RNA, there are four different bases, but thymine is found only in DNA and uracil is found only in RNA (i.e. in RNA, uracil replaces thymine).

All DNA (except some in viruses) is 'double-stranded'. This means that there are two polynucleotides, or 'strands', side by side. The bases on opposite strands are joined together by hydrogen bonds (Figure 25). These are not as strong as the covalent bonds joining adjacent nucleotides, but because there are so many it is quite difficult to break the strands apart. This helps to make DNA a very stable molecule (Figure 26).

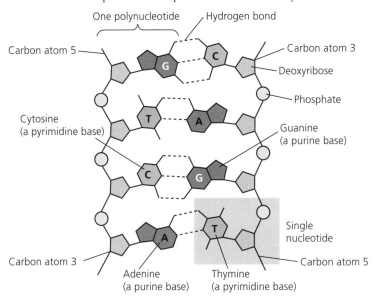

Note: Adenine (A) always bonds with thymine (T)
       Guanine (G) always bonds with cytosine (C)

Figure 25 The molecular structure of DNA shows two **antiparallel** polynucleotides. Notice that there are two hydrogen bonds between A and T and three between C and G

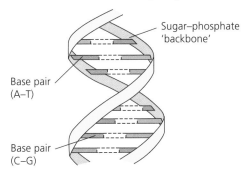

Figure 26 The two 'strands' are twisted around each other to form the famous DNA double helix (this shows a small section of DNA)

Base pairing in DNA is always as follows:
- adenine to thymine (A–T)
- guanine to cytosine (G–C)

The **antiparallel** arrangement of the polynucleotides in a molecule of DNA means that one polynucleotide is oriented in the 3' to 5' direction and the opposite polynucleotide in the 5' to 3' direction.

Molecules of RNA are 'single-stranded' polynucleotides. There are hydrogen bonds between some base pairs in tRNA to give a 'cloverleaf' shape. There are no hydrogen bonds in mRNA: it is a single, unfolded chain.

DNA and mRNA are compared in Table 12.

Table 11 The three types of RNA and their functions

Type of RNA	Function
Messenger RNA (mRNA)	Takes copies of genes from DNA in the nucleus to ribosomes
Ribosomal RNA (rRNA)	Helps make sites in ribosomes for assembling proteins from amino acids
Transfer RNA (tRNA)	Carries amino acids to ribosomes

Table 12 Comparing DNA and messenger RNA

Feature	DNA	Messenger RNA
Overall structure	Double helix	Single chain
Overall size of molecule	Very large	Small
Number of polynucleotide chains	2	1
Name of pentose (5C) sugar	Deoxyribose	Ribose
Nitrogenous bases	Adenine (A), thymine (T), cytosine (C) and guanine (G)	Adenine (A), uracil (U), cytosine (C) and guanine (G)
Base pairing	A pairs with T (A–T) C pairs with G (C–G)	No base pairing
Function	Long-term storage of genetic information	Transfer of genetic information from nucleus to ribosomes
Where found in eukaryotic cells	Nucleus (also some in mitochondria and chloroplasts)	Nucleus and cytoplasm (and in mitochondria and chloroplasts); also attached to ribosomes when protein is being made

DNA replication

Key concepts you must understand

DNA replication is DNA copying itself. Cells provide 'energised' nucleotides and enzymes for the process, but the important point is that DNA acts as a **template** so that new polynucleotide chains are built up on already existing ones. This is called **semi-conservative replication**, as the new DNA contains one 'old' polynucleotide (the template) and one 'new' polynucleotide.

Base-pairing is important because exposed bases on the template DNA determine which nucleotide is next in the sequence. Cytosine always pairs with guanine and adenine always pairs with thymine. Replication happens during interphase of the cell cycle (p. 57).

Exam tip

Conservative replication would involve producing DNA that had two newly synthesised polynucleotides, rather than one old and one new, as in semi-conservative replication.

Key facts you must know

Think of replication as a process that copies DNA very accurately, so minimising the number of 'copying errors'.

- The enzyme helicase unwinds the double helix of DNA and hydrogen bonds holding the bases together are broken.
- Polynucleotide chains separate, exposing bases along both polynucleotide chains.
- Each chain acts as a template so a new chain can be built up, following the rules of base pairing — A–T and C–G.
- Free 'energised' nucleotide molecules in the nucleus are put in position alongside the exposed bases — each energised nucleotide consists of a pentose sugar, base and three phosphates (Figure 27).
- As the nucleotides 'line up', they form a growing chain.
- Two of the phosphates from each nucleotide break off in a reaction that forms a phosphodiester bond between the new nucleotide and the growing chain (Figure 28).
- The enzyme DNA polymerase joins the nucleotides in the correct sequence.
- DNA polymerase checks the new base sequence to make sure it is correct, effectively 'proofreading' the new polynucleotide. If there is an incorrect nucleotide, it cuts it out and replaces it.
- Hydrogen bonds form between the bases on opposite polynucleotide chains — the template chain and the newly synthesised chain.
- DNA winds up again into a double helix. Replication is complete.

Although DNA polymerase proofreads the polynucleotide chain as it makes it, not everything gets corrected. There are errors — these random changes to the sequence of bases in DNA are spontaneous mutations. In humans it has been estimated that there are between 70 and 130 such mutations in each person in each generation.

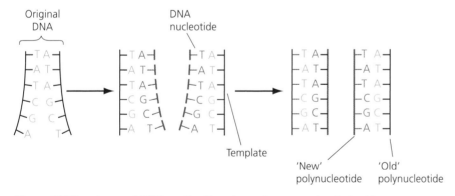

Figure 27 The stages of DNA replication shown in a very simple fashion

Knowledge check 16

Both polynucleotides act as templates during DNA replication. Explain the term *template* in this context.

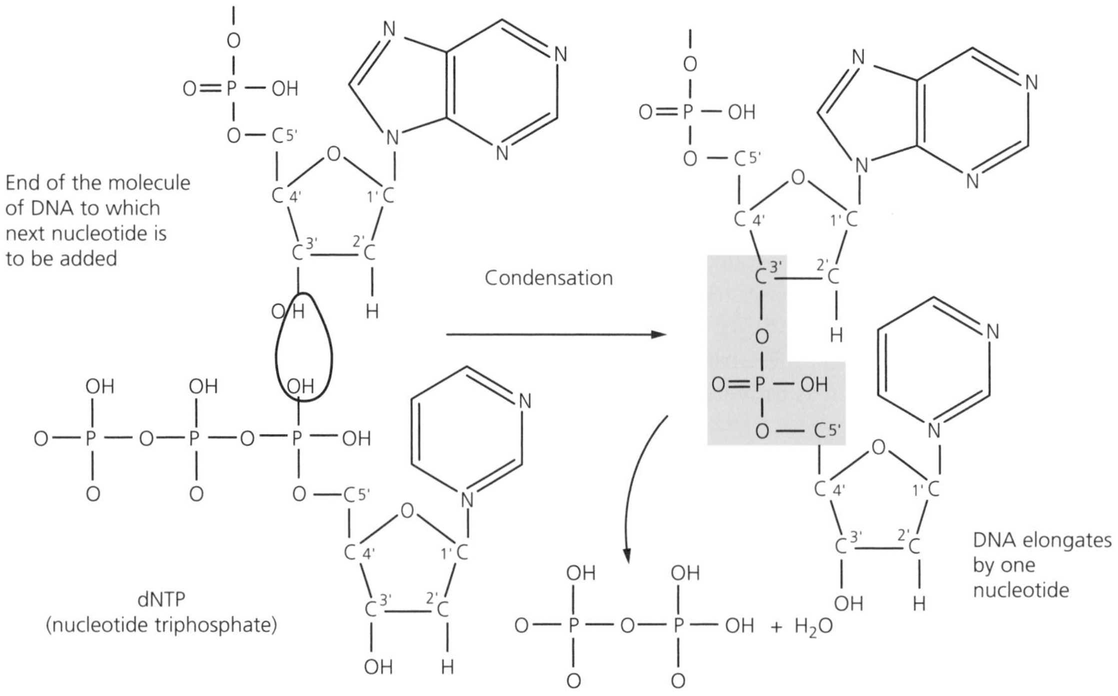

Figure 28 The formation of a phosphodiester bond. In DNA and RNA the phosphodiester bond forms between the 3′ carbon atom of one sugar molecule and the 5′ carbon atom of another, as shown by the yellow shading

'Energised' nucleotides

'Energised' nucleotides have three phosphate groups and are phosphorylated; they are more properly known as nucleoside triphosphates (NTPs). In DNA replication the sugar is deoxyribose, so these are dNTPs and there are four of them: dATP, dGTP, dCTP and dTTP. RNA is made from four rNTPs which contain ribose rather than deoxyribose. rATP fulfils other roles in cells including acting as the universal 'energy currency'.

During energy-consuming processes, such as protein synthesis and active transport, the hydrolysis of rATP, which is usually just known as **ATP**, provides energy when it is hydrolysed to form **ADP** and the phosphate ion, often called inorganic phosphate (P_i for short). During energy-generating processes of respiration and photosynthesis energy is used to reform ATP from ADP and P_i.

Links

Decoding the human and other genomes has involved 'reading' sequences of bases throughout DNA. The way in which genomes are sequenced is covered in Module 6 (Student Guide 4 in this series). The way in which information about base sequences is used is covered in Module 4 (Student Guide 2). The common energy currency, ATP, pops up everywhere. Here you need to know about its structure as a phosphorylated nucleotide and that it is interconverted to ADP. The role of these two compounds in energy transfer in cells is covered in Module 5 (Student Guide 3).

ATP and **ADP** are phosphorylated nucleotides. ATP (rATP) is formed from ADP (rADP) and phosphate. ATP is the way energy is transferred in cells between energy-generating processes and energy-consuming processes.

Knowledge check 17

a Use Figure 24 to make a sketch showing the structure of ATP.
b Write a simple equation to show the relationship between ATP and ADP.

Key facts you must know

One gene: one polypeptide

A gene determines the sequence of amino acids in a polypeptide. Each haemoglobin molecule is made from four polypeptides — two α-globin polypeptides and two β-globin polypeptides.

In humans, the genes for these two polypeptides are on different chromosomes. Each gene is a specific sequence of the four bases (A, C, T and G), which code for 20 different amino acids. The bases are 'read' in groups of three. Some of these triplets are given in Table 13. There are 64 of these triplets in total and together they form the **genetic code**. The code is described as **universal** as it is found in all organisms and in viruses.

The **genetic code** consists of triplets of bases in nucleic acids. Each amino acid is coded by at least one triplet of bases.

Table 13 Genetic dictionary: DNA triplet codes for three amino acids on the coding and template polynucleotides

Amino acid	Triplets on coding polynucleotide of DNA	Triplets on template polynucleotide of DNA
Glycine	GGT; GGG; GGC; GGA	CCA; CCC; CCG; CCT
Valine	GTT; GTG; GTC; GTA	CAA; CAC; CAG; CAT
Cysteine	TGT; TGC	ACA; ACG

The genetic code is described as **degenerate** because there are more codes than are needed for each amino acid. This is an advantage because the codes for each amino acid usually differ in the third base of the triplet, which reduces the effects of mutation. There are three 'stop' codes that do not code for amino acids. They indicate the end of a sequence (e.g. TGA on the coding strand of DNA). The triplet code for the amino acids is called the genetic code and is shown in a genetic dictionary as the RNA code or as one of the DNA codes. Each group of three bases in mRNA that codes for an amino acid is known as a **codon**.

Knowledge check 18

State the DNA triplets for histidine (His) and phenylalanine (Phe).

Knowledge check 19

Explain why the code for amino acids must be a triplet code and not a two-base code.

Exam tip

The complete genetic code can be found on many websites. You are not expected to remember any of the triplets or codons, but you may have to use the genetic code in answering a question, in which case it will be provided for you. Do *not* confuse the genetic code (triplets of bases that code for each amino acid) with sequences of bases that code for specific polypeptides. The genetic code is not the sequence of bases.

Protein synthesis

The three stages in protein synthesis are **transcription**, amino acid activation and **translation**.

Transcription

β cells in the islets of Langerhans in the pancreas make insulin. It is only in these cells that the gene for insulin is switched on. Each β cell has two copies of this gene,

Transcription is the production of mRNA from a DNA template by RNA polymerase; the mRNA is a copy of a gene or genes.

Translation is the production of a polypeptide by ribosomes that assemble amino acids using the sequence of codons on mRNA.

but many copies are needed to send to the thousands of ribosomes in the cell to make the quantities of insulin required. Short-lived copies of the gene are made by transcription. The copies are molecules of mRNA.

The process of transcription is shown in Figure 29. Note that the base sequence of the mRNA is the same as that of the coding polynucleotide and complementary to that of the template polynucleotide (U replaces T in RNA).

> **Exam tip**
>
> Do not confuse transcription with translation. TransCription comes first and changes DNA code into RNA code. TransLation comes next and uses the RNA code to make a different 'language' made up of amino acids.

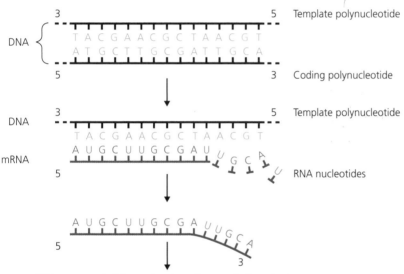

Figure 29 Transcription of the (non-coding) template polynucleotide in DNA

1. Hydrogen bonds between the bases in the two polynucleotide chains break in the area of DNA that corresponds to the insulin gene.
2. One polynucleotide acts as a template for the synthesis of mRNA.
3. Free 'energised' RNA nucleotides (rNTPs) in the nucleus pair up with the exposed bases on the template polynucleotide.
4. The nucleotides are joined together to form a polynucleotide — mRNA. This process is catalysed by the enzyme RNA polymerase.
5. mRNA leaves the nucleus through a nuclear pore.

Amino acid activation

Amino acids are 'identified' or 'tagged' by combining them with molecules of transfer RNA (tRNA). You could think of the nucleotide 'labels' as being similar to barcodes. The nucleotide 'barcode' is the **anticodon**. Enzymes in the cytoplasm attach amino

> **Exam tip**
>
> You should watch some animations of transcription and translation. Some of the best are on the websites of John Kyrk, Learn. Genetics and the DNA Learning Center.

> **Knowledge check 20**
>
> List the first five amino acids in the polypeptide coded for by the gene in Figure 29.

> **Knowledge check 21**
>
> What are the likely effects of reading errors in transcription?

> **Knowledge check 22**
>
> State what 3′ and 5′ signify.

acids to specific tRNA molecules. This is not a random process — each amino acid is identified by a specific tRNA molecule (Figure 30). The hydrolysis of ATP provides the energy for the process.

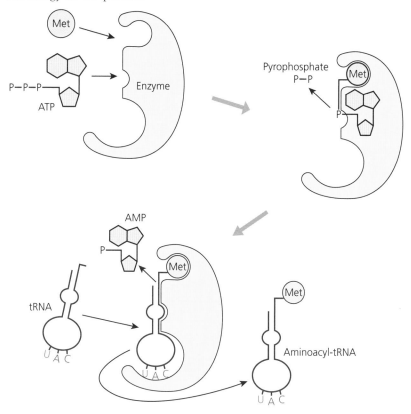

Figure 30 The enzyme shown here only accepts methionine and its specific tRNA molecule, with the anticodon UAC

Translation

1 The mRNA molecule joins with a ribosome in the cytoplasm.

2 Each ribosome has two sites to hold two tRNA molecules at the same time. Each tRNA molecule is attached to its specific amino acid.

3 Each tRNA molecule has a sequence of three bases (anticodon) that pairs with three bases (codon) on mRNA.

4 tRNA and mRNA pair together following the rules of complementary base pairing (A with U; C with G).

5 A condensation reaction occurs between the amino acids to form a peptide bond. This is catalysed by a molecule of rRNA known as a ribozyme.

6 The ribosome moves along the mRNA molecule 'reading' the sequence of bases. It reads these bases in groups of three that do not overlap. You can imagine this as a 'reading frame', which moves along the mRNA without any overlaps between the triplets.

Knowledge check 23

The genetic code is described as universal, degenerate and non-overlapping. Explain this description.

7   As this happens, a polypeptide grows by the addition of new amino acid molecules.

8   When an amino acid has joined to the growing chain, its tRNA molecule leaves the ribosome to attach to another amino acid (Figure 31).

9   When the ribosome reaches a stop codon, the polypeptide breaks away and begins to fold spontaneously into its secondary and tertiary structure.

10 The cell processes the polypeptide, perhaps by combining it with other polypeptides to form a protein with quaternary structure, as happens in the formation of haemoglobin. In β cells in the islets of Langerhans in the pancreas a long polypeptide is cut up into two smaller polypeptides that are joined together by disulfide bonds to give insulin its quaternary structure.

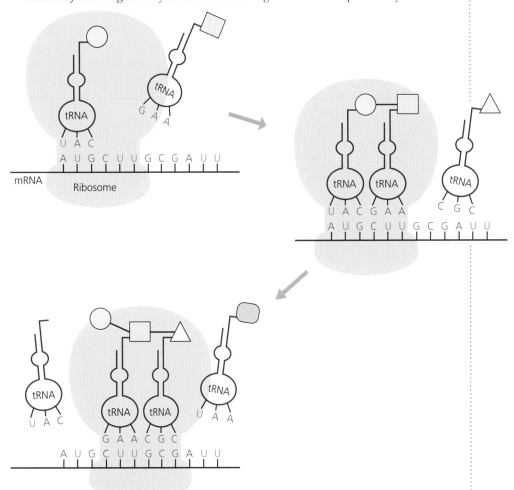

Figure 31 Translation

Summary

- Nucleic acids (RNA and DNA) are polymers of nucleotides. Each nucleotide consists of a pentose sugar, phosphate and a base. Genetic information is stored as DNA and retrieved in the form of mRNA during protein synthesis.
- DNA is a polynucleotide that is usually double-stranded. The nucleotides contain deoxyribose, phosphate, the purine bases adenine (A) and guanine (G), and the pyrimidine bases thymine (T) and cytosine (C).
- Hydrogen bonding between A and T and between G and C holds the two polynucleotides of DNA together. The polynucleotides are antiparallel, with carbons 3 and 5 of deoxyribose facing in opposite directions, giving a double helix with one strand running in a 3′ to 5′ direction and the other in a 5′ to 3′ direction.
- RNA is a polynucleotide that is usually single-stranded. The nucleotides contain A, G, C and U (uracil). There are three forms of RNA: messenger RNA (mRNA), transfer RNA (tRNA) and ribosomal RNA (rRNA).
- In semi-conservative replication, each polynucleotide of DNA acts as a template for the assembly of a new polynucleotide by DNA polymerase. The enzyme catalyses the formation of phosphodiester bonds between a free nucleotide and an elongating polynucleotide. The enzyme corrects mistakes in the base pairing between nucleotides and the template polynucleotide.
- Phosphodiester bonds are formed by a condensation reaction and broken by a hydrolysis reaction.
- A gene is a sequence of bases in DNA that codes for the assembly of amino acids to give the primary structure of a polypeptide.
- Genes code for the assembly of amino acids to make polypeptides, such as enzymes. Some enzymes, such as lysozyme, are composed of a single polypeptide; catalase is composed of four.
- The genetic code is the sequence of the four bases in DNA (A, T, C and G) that code for amino acids. There are 61 different triplets in DNA and codons in RNA that code for each of the 20 amino acids. There are three stop codons that do not code for any amino acids.
- Transcription is the copying of a nucleotide sequence in DNA into a complementary sequence in mRNA. Translation involves assembling amino acids on ribosomes using the sequence of codons to specify the sequence of amino acids in a polypeptide.
- Transfer RNA molecules are activated by combining with specific amino acids. The anticodons on tRNA molecules pair with codons on mRNA so that amino acids are assembled in the correct sequence. Peptide bonds form between amino acids.

■ Enzymes

Key concepts you must understand

Enzymes are proteins that catalyse chemical reactions. Without enzymes, these reactions would occur too slowly to support life as we know it. Reactions occur when molecules collide. Enzymes provide a place where reactions are likely to occur because they hold molecules under a strain, causing bonds to break and/or form. When molecules collide with enzymes in this way, they are described as successful collisions.

In **metabolism**, enzymes catalyse reactions that break down molecules to smaller products, for example the breakdown of hydrogen peroxide, and build up (synthesise) larger molecules from smaller ones, for example in the formation of macromolecules. They also move groups, such as phosphate groups, between compounds.

The synthesis of fibrous proteins (collagen, keratin and elastin) for animal tissues and cellulose for plant cell walls means that enzymes also have an important role in building the structure of organisms.

Key facts you must know

Enzymes are globular proteins with a tertiary structure held together by intramolecular bonds (Figure 22, p. 25). Different enzymes have different 3D shapes.

Enzymes:

- provide a site where molecules are brought together so that reactions occur more easily than elsewhere
- remain unchanged at the end of a reaction
- catalyse reactions in which compounds are built up
- catalyse reactions in which compounds are broken down
- change substrate molecules into product molecules

Intracellular enzymes work inside cells; examples are catalase and the enzymes of respiration and photosynthesis. **Extracellular enzymes** work outside cells; examples are amylase, lipase and trypsin, which are secreted into the small intestine for chemical digestion.

How enzymes work

The **active site** is the part of an enzyme where reactions occur. It is a cleft or depression on the surface of the molecule — a shape that fits around the substrate molecule. Enzymes and their substrates fit together like a **lock and key** to form an **enzyme–substrate complex**. When the reaction is complete this complex becomes an **enzyme–product complex** and then the product (or products) leaves so that another substrate molecule can enter the active site (Figures 32 and 33).

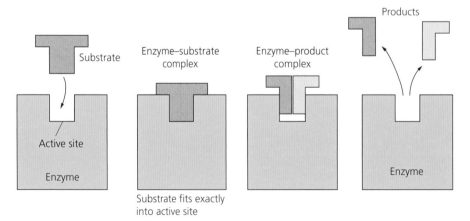

Figure 32 The lock-and-key mechanism of enzyme action

> **Metabolism** refers to the sum of all the chemical and physical changes that occur in an organism.

> **Exam tip**
>
> Enzymes function both inside cells (cellular level) and outside cells. Wherever enzymes function they influence the activity of the whole organism. You will find examples of this as you read on.

'Lock-and-key' is a simple model. Studies of enzymes using X-rays show that the **induced-fit** model, in which the enzyme's active site moulds around the substrate, is more likely (Figure 33).

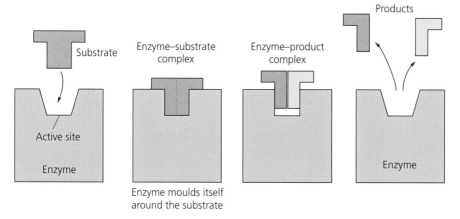

Figure 33 The induced-fit mechanism of enzyme action

Each enzyme has a specific shape and usually only one type of substrate molecule fits the active site. Note that the substrate is *not* the same shape as the active site. It has a shape that fits into the active site, i.e. the two have **complementary** shapes. Some enzymes are not as specific as others, having active sites that will accept a variety of substrate molecules with similar shapes.

Activation energy is the energy needed to be overcome before molecules can react (Figure 34). The reactions involved in breaking and making strong covalent bonds occur extremely slowly without enzymes because there is not enough energy to change substrate molecules to product molecules.

Enzymes provide a different route for the reaction to occur that has a lower activation energy. This allows reactions to occur that otherwise would need high temperatures and pressures to happen — conditions that are not compatible with life.

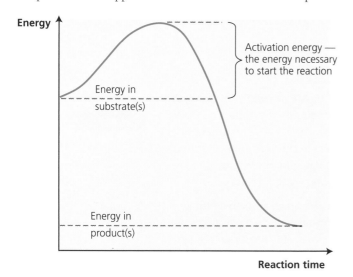

Figure 34 A graph showing the idea of activation energy

Exam tip

Complementary (*not* complimentary) refers to the matching shapes of an enzyme's active site and the substrate molecule that fits into it. **Specificity** applies to enzymes that act on one substrate, or a limited range of substrate molecules. Make sure to use both terms when explaining how enzymes work.

Enzyme molecules have specific shapes determined by their tertiary structure (p. 25). Amylase and trypsin are like this and have one active site per molecule. Some enzymes have quaternary structure, since their molecules comprise two or more polypeptides. The enzyme catalase has four identical polypeptides and thus has four active sites per molecule.

The tertiary structure of enzymes is altered by changes in pH and increases in temperature. As a result, active sites change shape and no longer accept substrate molecules.

Links

Many practical investigations involve enzymes. You need to have an understanding of what enzymes are and how they function to explain what happens in these experiments. Note that the active site is complementary in shape to the substrate. Other examples of this 'fitting together' are:

- antigens and antibodies (see Module 4 in Student Guide 2 in this series)
- antigens and receptors on the surfaces of T and B lymphocytes
- cell signalling: hormones (e.g. insulin) and their receptors; neurotransmitters and their receptors at synapses (see Module 5 in Student Guide 3)

Factors that influence enzyme activity

Key concepts you must understand

Enzymes are globular proteins. The active site is formed by folding of the protein (tertiary structure). Denaturation is the loss of shape of the active site when intramolecular bonds break at high temperature. Changes in pH can also cause them to break.

Rates of reaction are influenced by the initial concentration of substrate and by the concentration of enzymes. Many enzymes require cofactors in order to work. Inhibitors fit into enzyme molecules, preventing them from working.

Key facts you must know

Temperature

Separate test tubes of starch solution and amylase solution are placed in a water bath at 10°C and left to reach this temperature; the two are then mixed together. Samples are taken at intervals from the reaction mixture and tested with iodine solution, until there is no colour change. The time taken for the reaction to finish can be converted into a rate by calculating 1/(time taken). This is then repeated for more temperatures (e.g. 0, 20, 30, 40, 50 and 60°C), and the results plotted on a graph (Figure 35).

You can also follow the reaction using a colorimeter. Samples taken at the beginning of the reaction are dark blue and give a high reading for absorbance. Samples taken at the end of the reaction are light yellow and give a low reading. The initial rate can be calculated by drawing a graph of absorbance readings against time and taking a tangent.

Exam tip

Equilibration is the term used to describe leaving separate tubes of solutions of substrate and enzyme to reach the desired temperature *before* mixing them together.

Exam tip

Enzyme activity is sometimes used as the label for the y-axis instead of rate of reaction.

It is better to take the absorbance readings of known concentrations of starch and iodine solution and draw a calibration curve (absorbance plotted against concentration of starch). The calibration curve allows you to convert absorbance readings taken as the reaction proceeds into concentrations of starch.

Note that you cannot add iodine solution to the reaction mixture and watch it gradually change colour because iodine inhibits amylase.

Exam tip

Note that the graphs on the next few pages do not have plotted points. These graphs show the expected trends, not actual experimental results.

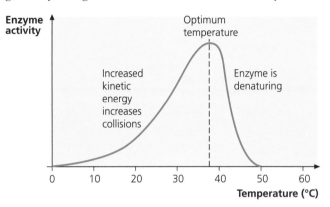

Figure 35 The effect of temperature on the rate of an enzyme-catalysed reaction

This is a *description* of the effect of temperature, as shown in Figure 35:
- Enzymes show no activity at freezing temperatures.
- Their activity increases as they are warmed from freezing.
- They are most active at their optimum temperature (for human enzymes this is 37°C).
- They are less active at temperatures below and above the optimum.
- There is no activity at high temperatures (e.g. above 50°C).

This is an *explanation* of these effects:
- At freezing temperatures there is no molecular movement, so no collisions occur.
- Enzymes are not denatured by freezing as they function on warming up.
- As temperature increases, substrate and enzyme molecules have more kinetic energy, so that there are more successful collisions between substrate molecules and enzyme molecules to form enzyme–substrate complexes.
- At high temperatures, there is excessive movement within the enzyme molecules so bonds (e.g. ionic and hydrogen) break. The active site changes shape and no longer accepts the substrate — the enzyme molecules are denatured.

Exam tip

Not all enzymes are like this. Enzymes from organisms that live in very cold habitats have optimum temperatures much lower than 37°C. Enzymes from bacteria that live in hot springs are active at temperatures up to 90°C.

🖩 The **temperature coefficient** is the ratio between the rate of a reaction at two different temperatures, usually 10°C apart. The symbol Q_{10} is used to represent this ratio. The Q_{10} for enzyme-catalysed reactions up to the optimum temperature is usually 2, i.e. the rate of reaction doubles for each increase of 10°C.

The formula is:

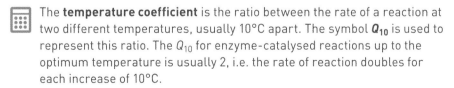

$$Q_{10} = \frac{\text{rate of reaction at } x + 10°C}{\text{rate of reaction at } x°C}$$

pH

An investigation similar to the one described for temperature can be carried out for pH using different buffer solutions with the reaction mixtures. Buffer solutions maintain a constant pH. There are different buffer solutions available to give a range between pH 3 and pH 11. The results of such an investigation are shown in Figure 36.

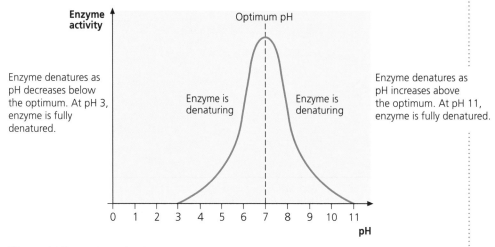

Figure 36 The effect of pH on an enzyme-catalysed reaction

This is a *description* of these effects:

- The enzyme is most active at a certain pH — the optimum pH.
- It is less active either side of the optimum pH.
- It is inactive at extremes of pH.

This is an *explanation* of these effects:

- pH is a measure of hydrogen ion concentration; as the pH of a solution changes, the charges on the R groups of amino acids change.
- At low pH, when the concentration of hydrogen ions is high, many of these ions interact with negatively charged R groups, so cancelling out their charge.
- Similar changes occur when the pH increases.
- Changes in hydrogen ion concentration disrupt the ionic bonding between oppositely charged R groups within the tertiary structure — the enzyme shape changes.
- When active sites change shape, they no longer have shapes complementary to their substrates and cannot form enzyme–substrate complexes.

Substrate concentration

The effect of substrate concentration is investigated by setting up a series of test tubes, all with the same concentration of enzyme, but with different concentrations of substrate. The initial rate of reaction in each tube is determined and plotted on a graph (Figure 37).

Exam tip

pH is not a simple linear scale. A change in one pH unit (e.g. pH 7.0 to 8.0) represents a change in concentration of hydrogen ions by a factor of 10. The scale is therefore logarithmic.

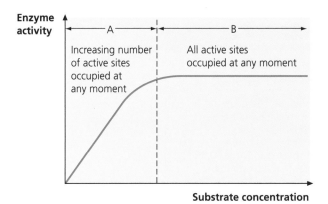

Figure 37 The effect of substrate concentration on enzyme activity

As the concentration of substrate is increased, the rate of reaction increases (region A on the curve) because there are more substrate molecules for the enzymes to act on. Substrate concentration is the factor that **limits** the rate of reaction. At high concentrations of substrate (region B), all the active sites are filled and enzyme activity is limited by the enzyme concentration.

Enzyme concentration

The investigation is repeated with the substrate concentration kept constant and the enzyme concentration increased. The result is shown in Figure 38.

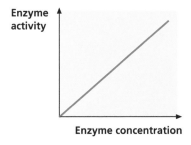

Figure 38 The effect of enzyme concentration on enzyme activity

With more enzyme molecules, there are more active sites available and so the only limiting factor is the enzyme concentration. This happens if there is an excess of substrate at each enzyme concentration tested.

Cofactors and coenzymes

Some enzymes do not function unless they are combined with a **cofactor**. Cofactors can be ions or complex organic substances, which may occupy the active site and take part in the reaction or are involved in other ways. **Coenzymes** are organic cofactors that are changed slightly during the reaction catalysed by the enzyme. Most coenzymes are mobile and travel back and forth between enzymes. Table 14 lists the cofactors and coenzymes required by some enzymes.

Exam tip

When describing graphs like those in this section, make sure that you use words such as increase, decrease, peak, maximum, minimum, constant and plateau. It is often a good idea to divide graphs into sections (e.g. A, B, C...) and then write about each.

A **cofactor** is any ion or organic compound that is required during an enzyme-catalysed reaction but is not the substrate of the reaction.

A **coenzyme** is any organic cofactor, for example coenzyme A.

Table 14 Cofactors and coenzymes

Enzyme	Cofactor	Role of enzyme
Amylase	Cl^-	Hydrolyses glycosidic bonds in starch
Urease	Nickel (Ni^{2+})	Hydrolyses urea to ammonia and carbon dioxide
Carbonic anhydrase	Zinc (Zn^{2+}) as a prosthetic group	Involved in transport of carbon dioxide in the blood (see Module 3 in Student Guide 2)
Catalase	Haem (containing iron as Fe^{2+})	Breaks down hydrogen peroxide
Enzyme(s)	**Coenzyme**	**Role of coenzyme**
Several enzymes in photosynthesis	NADP	Transfer of hydrogen from water to carbon dioxide in photosynthesis
Several enzymes in respiration	NAD and FAD	Transfer of hydrogen during respiration
Pyruvate decarboxylase (a mitochondrial enzyme)	Coenzyme A	Transfers acetyl groups (C_2H_3O) from pyruvate and fatty acids during respiration

Many coenzymes are derived from vitamins — for example, NAD and NADP are derived from nicotinic acid (vitamin B_3), and FAD from riboflavin (B_2). Pantothenic acid is part of coenzyme A. Thiamine (B_1) is required for the conversion of pyruvate into coenzyme A.

Inhibitors

Some enzyme inhibitors attach permanently to enzyme molecules and stop the enzyme functioning completely. These are **non-reversible inhibitors** so the only way for a cell to overcome the inhibition is by synthesising new molecules of the enzyme. **Reversible inhibitors** attach to enzymes temporarily; they act to slow down enzyme-catalysed reactions by fitting into sites on the enzyme. There are two types of reversible inhibitor.

Competitive inhibitors:
- fit into the active site
- have a shape similar to, but not the same as, the substrate
- block the substrate from entering the active site
- prevent the formation of enzyme–substrate complexes
- have an effect that can be reversed by increasing the concentration of substrate

The mode of action of a competitive inhibitor is shown in Figure 39.

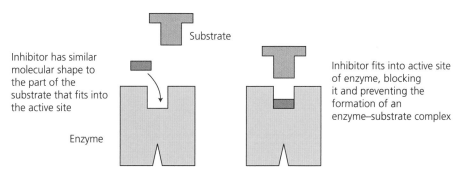

Figure 39 Competitive inhibitors fit into the active site and block it

Non-competitive inhibitors:

- do not fit into the active site
- fit into a site elsewhere on the enzyme
- cause the active site to change shape so it is no longer complementary in shape to the substrate
- have an effect that cannot be reversed by increasing the concentration of substrate

The mode of action of a non-competitive inhibitor is shown in Figure 40.

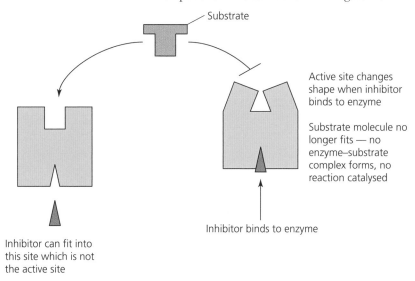

Figure 40 Non-competitive inhibitors fit into another site on the enzyme, leading to a change in the active site

Non-competitive inhibition is used to control multi-step metabolic pathways in cells. Each reaction in a pathway is catalysed by a different enzyme. As the final product accumulates, it inhibits the enzyme at the beginning of the pathway, so slowing down or stopping production. This is called **end-product inhibition** and is used to ensure that cells do not waste energy and resources making products that are not required. Some inhibitors and the enzymes that they inhibit are listed in Table 15.

Table 15 Competitive and non-competitive inhibitors and the enzymes that they inhibit

Enzyme	Competitive inhibitor	Role of enzyme
Succinate dehydrogenase (a mitochondrial enzyme)	Malonic acid	Transfers hydrogen during respiration
HMG-CoA reductase	Statins, e.g. rosuvastatin	Involved with the synthesis of cholesterol in the liver
Transpeptidase in some bacteria	Penicillin*	Cell wall synthesis in some bacteria
Urease	Thiourea	Hydrolyses urea to ammonia and carbon dioxide
Enzyme	**Non-competitive inhibitor**	**Role of enzyme**
Cytochrome oxidase (a mitochondrial enzyme)	Potassium cyanide KCN*	Catalyses an important step in respiration in mitochondria

Asterisks (*) indicate non-reversible inhibitors. The others are reversible.

Many drugs act on enzymes to inhibit them; examples include penicillin and the statins (see Table 15). Some poisons are enzyme inhibitors. If you see an examination question that tells you that the rate of an enzyme-catalysed reaction slows down following the addition of a substance, then that substance is likely to be an inhibitor. If the effect is reversed by increasing the concentration of substrate, then it is a competitive inhibitor.

If enzymes are inhibited permanently, the only way a cell can overcome this inhibition is to produce more enzymes by protein synthesis. This takes time and will not happen if the inhibitor is fatal.

Links

As part of your practical work you may investigate the effect of a factor on the activity of an enzyme. You are expected to explain how you would keep control variables constant, for example by using a buffer solution to keep pH constant if investigating temperature. See Question 9 of the AS-style paper on p. 77.

You will learn more about coenzymes in respiration and photosynthesis in Module 5. NAD and NADP are mobile coenzymes that move between enzymes in the cell, transferring hydrogen between compounds. FAD is bound permanently as a prosthetic group to the enzyme succinate dehydrogenase, which is situated in the inner mitochondrial membrane.

Exam tip

This table may look difficult because of all the long names. It shows you the importance of enzyme inhibitors. Make sure that you learn the principles of enzyme inhibition and can explain them (see the summary on p. 49).

Summary

- Enzymes are globular proteins that catalyse metabolic reactions involving the breakdown of substrate molecules (e.g. hydrogen peroxide) and synthesis of molecules (e.g. DNA). Intracellular enzymes act inside cells; extracellular enzymes work outside cells, for example in the gut lumen.
- Activation energy needs to be overcome before a reaction can proceed. Enzymes effectively lower activation energy by having active sites where substrate molecules fit either because they have a complementary shape (lock and key) or because the enzyme moulds around the substrate (induced fit) to form an enzyme–substrate complex.
- Enzymes have different degrees of specificity. Some only accept one type of substrate molecule; less specific enzymes accept several similar substrate molecules.
- Rates of activity are influenced by factors that include temperature, pH, substrate concentration and enzyme concentration. The effect of each factor is investigated by using five or more reaction mixtures across an appropriate range (e.g. 0°C to 70°C) and determining the initial rate of reaction and plotting a graph. The effect of each factor gives a characteristic trend or pattern.
- Cofactors are required for the functioning of enzymes. Ions, such as Zn^{2+}, fit into active sites and take part in the reaction but are not changed like a substrate. Coenzymes are complex organic cofactors. Some of these are reduced or oxidised during reactions and act to transport hydrogen between compounds.
- Non-reversible inhibitors combine permanently with enzymes and stop them working. Reversible inhibitors reduce the activity of enzymes by either competing with the substrate for the active site (competitive inhibitors) or attaching to another site on an enzyme to alter the shape of the active site (non-competitive inhibitors). Poisons (e.g. cyanide) and medicinal drugs (e.g. penicillin) are enzyme inhibitors. Potassium cyanide inhibits cytochrome oxidase in mitochondria, so stopping respiration.

Biological membranes

Fluid mosaic structure of membranes

Key concepts you must understand

The roles of membranes are to form boundaries and to divide cells into compartments. The plasma membrane (also known as the cell surface membrane) forms the outermost boundary of the cell. This allows the composition of cells to be different from that of their external environment.

Membranes are barriers between the cytoplasm and the outside world, keeping in large molecules such as enzymes, RNA and DNA, and keeping out many others. Cells need to exchange substances with their surroundings, so membranes are permeable — not freely permeable to anything and everything, but **partially permeable** to some substances.

Organelles, for example mitochondria, chloroplasts, endoplasmic reticulum and Golgi apparatus, are made of membranes, and are separate compartments within cells. The lysosome membrane encloses enzymes and stops them breaking down molecules, such as proteins, in the cytoplasm. The cell surface membrane and some intracellular membranes form one continuous system because membrane from the Golgi apparatus becomes part of the cell surface membrane during exocytosis (Figure 43).

Key facts you must know

All membranes have the same basic structure — the fluid mosaic structure (Figure 41).

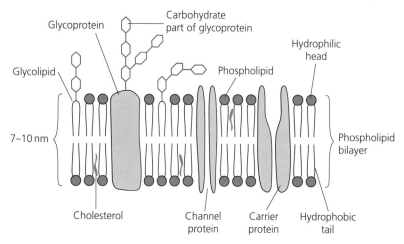

Figure 41 The fluid mosaic model of membrane structure

Figure 41 shows a cross-section of a tiny part of a membrane. It is composed of a double layer (bilayer) of phospholipids, together with proteins. Each phospholipid molecule has a 'head' and two 'tails'. The head end is polar and hydrophilic. The tails are non-polar and hydrophobic. Phospholipid heads are soluble in water; the tails provide a hydrophobic barrier that many water-soluble substances cannot cross easily. This hydrophobic barrier restricts the movement of substances in and out of cells, so helping to keep a constant environment inside the cytoplasm.

How are the components arranged?

Phospholipids

- Membranes have two layers of phospholipid, forming a bilayer.
- The molecules in the two layers have opposite orientations, so that the non-polar 'tails' associate with each other and the polar 'heads' face the cytoplasm and the fluid outside the cell.
- The polar 'heads' form hydrogen bonds with water in the cytoplasm and extracellular fluid.

Proteins

- Membrane proteins are embedded in the phospholipid bilayer.
- Transmembrane proteins extend right through the bilayer with one end in the cytoplasm and the other end extending to the outside. These are held in the membrane because they have hydrophobic regions that span the hydrophobic interior of the membrane.
- Some of the proteins in membranes are enzymes and catalyse reactions, such as those of photosynthesis in chloroplasts and respiration in mitochondria.

Exam tip

The carbohydrate chains are on the external surface of the membrane. This helps you to orientate diagrams of the plasma (cell surface) membrane if you are not told.

Exam tip

Look at Figure 17 (p. 22) to see more detail of phospholipids.

Exam tip

Proteins are not free to move all over the surface of some cells. Instead they are anchored in place by the cytoskeleton. This happens in cells that face two different environments, for example those that absorb food from the gut and pass it into the blood.

Carbohydrates

- These are short-branched chains of sugar molecules that are 'tree-like' attachments to proteins and lipids.
- Glycolipids are lipids with chains of sugar molecules attached.
- Glycoproteins are proteins with chains of sugar molecules attached.
- Carbohydrates are attached to lipids and proteins only on the external surfaces of cell membranes.

Cholesterol

- Cholesterol molecules have polar and non-polar regions. Polar regions bind to polar heads of phospholipids; non-polar regions bind to phospholipid tails.
- It maintains the stability of membranes by preventing phospholipids solidifying at low temperatures and becoming too fluid at high temperatures.
- Cholesterol reduces the permeability of membranes to water, ions and polar molecules.
- It is not found in the membranes of prokaryotes.

Why fluid mosaic?

Fluid

The membrane is held together mainly by hydrophobic interactions between the phospholipids and between proteins and phospholipids. These weak interactions allow the molecules to move so that the membrane is liquid. Phospholipid molecules move laterally in the plane of the membrane. Proteins are much larger and move more slowly — imagine protein molecules moving about like icebergs in a 'sea' of lipid.

Mosaic

A membrane is like a collage of many different proteins in the lipid bilayer. Think of a Roman mosaic that is made of tiny pieces of tile. Now think of the pieces constantly moving about and you should have a picture in your mind of a fluid mosaic.

Knowledge check 24

Explain the term *fluid mosaic* as applied to membranes.

Cell signalling

Many glycoproteins are receptors for chemical signals sent between cells. You will remember from GCSE that neurones (nerve cells) release chemicals into gaps known as synapses. These chemicals act as signals from one neurone to the next so information can be sent through the nervous system. You will also remember that some cells release hormones into the bloodstream as another type of signalling.

The release of signalling compounds by cells to influence other cells in the immediate surroundings is a third method. Cells called mast cells are an example. When damaged they release histamine, which signals to cells lining blood capillaries to become 'leaky' and allow more fluid to pass out from the blood.

The cells that receive such signals are called target cells, and they have receptors on the surface of their membranes that bind with the signalling molecule. The shape of the receptor is complementary to the shape of the signalling molecule so the two fit together.

It is possible to design drugs to bind to these receptors. For example, some asthmatics use salbutamol (Ventolin™) in an inhaler. This drug binds to receptors for adrenaline

Knowledge check 25

Explain how cells send and receive siginals.

on smooth muscle in the bronchioles, making these air passages widen, thereby making it easier to breathe.

Links

Membranes are involved in all exchanges between living things and their environment, for example, across alveoli in the lungs and across plant root hairs (see Module 3 in Student Guide 2). You will need to know the structure of membranes if asked to explain the effect of different temperatures on membrane permeability.

Cell signalling is a theme that recurs throughout the course. Note the importance of glycoprotein receptors as the 'receivers' on the target cells. Another theme that recurs is that of protein shapes. Receptors and signalling molecules fit together because they have complementary shapes in the same way that substrates and enzymes fit together.

Movement across membranes

Key concepts you must understand

Membranes are barriers, but they allow considerable exchange of substances between the cytoplasm and the surroundings. Some substances are small enough to pass through membranes quite easily; others are larger and need special methods. Some molecules move through the membrane down a concentration gradient, for example oxygen moves into animal cells and carbon dioxide diffuses out of animal cells in this way. This is **passive transport** as the cell does not use any of its energy from respiration to move the molecules.

All cells have a *lower* concentration of sodium ions than their surroundings, because they use membrane proteins to pump sodium ions out of the cells, utilising ATP to do this. This type of movement is **active transport**.

Some molecules and even quite large particles are moved into or out of cells surrounded by membrane. This type of movement is **bulk transport**, which also requires ATP from the cell to move vacuoles or small vesicles away from or towards the plasma membrane.

Key facts you must know

There are five ways in which substances can cross membranes, divided into two categories:

- Passive transport — not requiring energy from cells:
  - simple diffusion (Figure 42)
  - facilitated diffusion (Figure 42)
  - osmosis
- Transport mechanisms requiring energy from cells:
  - active transport (Figure 42)
  - bulk transport — endocytosis and exocytosis (Figure 42)

Exam tip

You can write: 'oxygen diffuses down its concentration gradient into the cell' or 'oxygen diffuses into the cell from a high concentration to a low concentration'. A gradient exists between the outside and inside of the membrane, so do not write '...from a high concentration gradient to a low concentration gradient'.

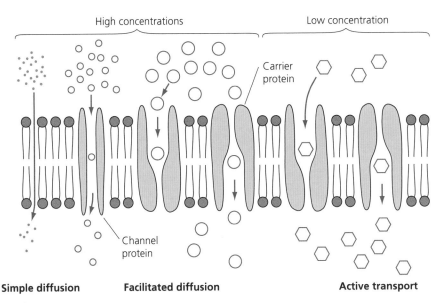

Figure 42 Diffusion and active transport

Simple diffusion

Non-polar molecules, such as steroid hormones (e.g. testosterone and oestrogen), lipid-soluble vitamins (e.g. vitamins A and D), many narcotics, the respiratory gases oxygen and carbon dioxide, and small polar molecules such as water and urea, move through the phospholipid bilayer down their concentration gradients. This is a passive process. Ions and polar molecules, such as sugars, cannot diffuse across membranes in this way.

Facilitated diffusion

Proteins play a key role in regulating transport across membranes. **Channel proteins** each have a hollow core, which acts as a water-filled channel or pore. This allows small, polar molecules and ions to diffuse across membranes. This is known as **facilitated diffusion** as the channels allow this to happen (facilitate = make easier).

Carrier proteins work by binding to the substance and physically moving it across. The binding causes a change in the shape of the carrier and results in the bound substance being released at the other side of the membrane. This is also a type of facilitated diffusion, when the substances involved move down their concentration gradients.

Osmosis

Osmosis is the diffusion of water across membranes. Water diffuses through the phospholipid bilayer and through special channel proteins, known as aquaporins. The direction in which water diffuses depends on **water potential** gradients, which are determined partly by the solute concentration in the cytoplasm and the external surroundings and partly by other factors, for example the pressure exerted by plant cell walls on the cytoplasm and vacuole.

Knowledge check 26

Distinguish between facilitated diffusion and simple diffusion. (Distinguish means give one or more differences.)

Osmosis is the diffusion of water down a water potential gradient through a partially permeable membrane.

Water potential is a measure of the ability of water to move from one place to another from a high water potential to a low water potential.

Active transport

Active transport involves carrier proteins. Cells use ATP as the source of energy to move substances from a low to a high concentration against a concentration gradient.

As mentioned on p. 52, many cells use active transport to move sodium ions out of the cytoplasm in exchange for potassium ions — the 'sodium pump'. This helps regulate their volume, as three sodium ions are pumped out for every two potassium ions pumped in. The net loss of one sodium ion for every molecule of ATP helps to prevent the cell water potential falling very low and too much water diffusing in by osmosis.

Bulk transport

Bulk transport (Figure 43) is used for the transport of larger molecules and particles. Exocytosis and endocytosis (pinocytosis and phagocytosis) are examples.

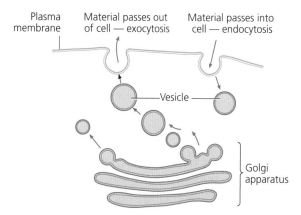

Figure 43 Bulk transport

Exocytosis

Substances are moved out of cells. Vesicles travel towards the cell surface and fuse with the membrane to extrude substances. For example, this happens in pancreatic cells that secrete insulin (see Question 6 of the A-level-style questions on page 90).

Endocytosis

Substances are brought into cells. Vacuoles or vesicles form at the cell surface and move into the cytoplasm, as happens when phagocytes engulf bacteria. Phagocytosis is the uptake of particles from the surroundings; pinocytosis is the uptake of liquids.

> **Exam tip**
>
> Endocytosis and exocytosis both require ATP from the cell, as these are energy-requiring processes. Be careful not to confuse them with active transport, which involves movement of substances through carrier proteins, not surrounded by membranes.

> **Exam tip**
>
> Vacuoles or vesicles? Both are surrounded by a single layer of membrane that isolates the contents from the surrounding cytoplasm. Vesicles are small vacuoles.

Summary

- Membranes are found at the surfaces of all cells and they make up some of the organelles in eukaryotic cells. They have a fluid mosaic structure comprising a phospholipid bilayer with proteins.
- Membranes are partially permeable in that they allow some substances to pass through but not others.
- Plasma membranes provide a barrier between the cytoplasm and the surroundings; they permit the entry of some substances through channel and carrier proteins; they allow for recognition of signalling compounds, such as hormones, in cell communication; they provide a site for enzymes.
- Intracellular membranes divide the cytoplasm of eukaryotic cells into compartments (organelles). Some of these membranes have enzyme in them to catalyse reactions, such as some in photosynthesis and respiration.
- Phospholipids have hydrophobic 'tails' that form the hydrophobic core of membranes; the polar 'heads' form hydrogen bonds with water in the cytoplasm and surrounding fluid. Non-polar substances pass through the bilayer by simple diffusion.
- Glycoproteins and glycolipids have carbohydrate chains to give them specific shapes to act as receptors for cell signalling compounds such as neurotransmitters and hormones. Drugs also bind to them.
- An increase in temperature and the presence of organic solvents, such as ethanol, disrupt membrane structure and so increase permeability.
- Channel and carrier proteins allow polar and charged substances to pass in and out of cells down their concentration gradients. This is facilitated diffusion.
- Active transport is the movement of a substance across a membrane against its concentration gradient by carrier proteins. Bulk transport is the movement of substances across a membrane within a vacuole. Both forms of transport require energy from the cell.
- Osmosis is a special type of diffusion in which water moves down a water potential gradient across a partially permeable membrane. Most water moves through aquaporins.
- The diffusion of water into and out of animal and plant cells can be investigated by immersing tissues in solutions of different water potential. The direction of the net flow of water is determined by the water potential gradient. If there is no change in mass of plant tissue, then there is no net diffusion and the water potential of the solution is the same as the water potential of the cells in the tissue.

Cell division, cell diversity and cellular organisation

The cell cycle

Key concepts you must understand

Multicellular organisms, such as animals and plants, grow in two basic ways: the cells increase in size and then they divide. Cells increase in size by making new molecules, such as phospholipids and proteins, new membranes and new organelles.

Cells cannot grow like this indefinitely. When they reach a certain size, diffusion distances between the plasma membrane and the centre of the cell become too great and not enough oxygen reaches the mitochondria for respiration. Also, there is not a large enough surface for sufficient diffusion of oxygen and carbon dioxide to occur relative to the size of the cell (see Module 4 in Student Guide 2). Before these problems arise, a cell divides into two daughter cells.

Multicellular organisms have groups of unspecialised cells in different parts of the body that retain the ability to divide, known as **stem cells**. Many cells, however, lose this ability as they develop into specialised cells. Some have very short lives, such as those lining the human gut, and so need to be replaced at a rapid rate. Their replacement is the function of stem cells. Stem cells are found throughout the bodies of animals although they are often concentrated in areas such as the skin, the lining of the gut and the bone marrow, where they produce replacement cells.

In plants, meristematic cells retain the ability to grow and divide and are the equivalent of stem cells. **Meristems** are areas where these cells are found, including root tips, shoot tips and the cambium that gives rise to xylem and phloem tissues.

Stem cells continuously grow and divide to produce daughter cells that replace dead or worn-out cells. The cycle of changes that they go through is the **cell cycle**.

Stem cells are the subject of much research. Scientists use stem cells to study the processes of cell growth and division to find out how these are controlled. Stem cell therapy is the use of these unspecialised cells to treat human and animal conditions in which cells have died and need replacing. For example, stem cells are used in the treatment of tendon, ligament and joint diseases in horses. Stem cells are removed from a horse, grown in culture and then implanted in appropriate places in the same horse.

The only commonly available stem cell therapy for humans is bone marrow transplants, in which stem cells for blood cells are inserted into a person who has a disorder such as leukaemia. Stem cell therapies may become available for heart and eye diseases and those involving death of central nervous tissue, such as Parkinson's and Alzheimer's diseases.

Key facts you must know

The cell cycle shows the different stages of a cell's life (Figure 44). During the first growth stage known as G_1 the cell synthesises proteins, makes organelles, such as mitochondria, grows larger and stores energy. The 'energised' nucleotides needed for DNA synthesis and amino acids for protein synthesis are made. DNA replication occurs during the S (synthesis) stage. The cell prepares for division in G_2 by making more proteins and storing more energy. Mitosis follows G_2 and is the division of the nucleus; in many dividing cells mitosis is followed by division of the cytoplasm during cytokinesis.

The cell cycle is controlled by a series of **checkpoints** that occur at intervals during interphase and mitosis. Each checkpoint ensures that the preceding stage is completed so that the cycle can move on to the next stage. There are three checkpoints that control the cycle. These are:

- G_1 checkpoint, which determines whether the cell goes through the cell cycle or does not divide
- G_2 checkpoint, which determines whether the cell is ready to enter mitosis
- metaphase checkpoint, which determines whether all the chromosomes are attached to the spindle ready for anaphase (see Figure 45)

Stem cells are undifferentiated cells that retain the ability to divide throughout an organism's life.

Meristems are areas of growth in plants where stem cells (meristematic cells) are concentrated.

The **cell cycle** is a sequence of changes that occurs during the life of a cell, including growth, synthesis of DNA, nuclear division by mitosis and division of the cytoplasm.

Exam tip

Interphase is *not* a resting phase. G (as in G_1 and G_2) stands for gap.

A **checkpoint** is a point during the cell cycle in which the cell assesses whether it is ready to progress to the next stage.

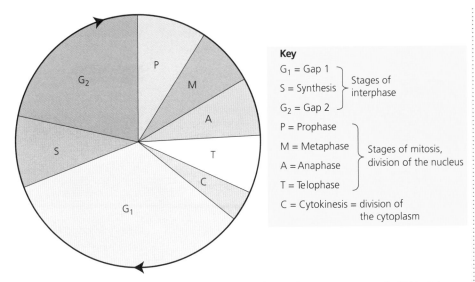

Key

G_1 = Gap 1
S = Synthesis   } Stages of interphase
G_2 = Gap 2

P = Prophase
M = Metaphase
A = Anaphase   } Stages of mitosis, division of the nucleus
T = Telophase

C = Cytokinesis = division of the cytoplasm

Figure 44 The cell cycle. In most cases, mitosis (PMAT) takes up about 5–10% of the length of time of each cycle, which is too short a time relative to interphase to show accurately on this diagram

In the metaphase checkpoint, centromeres that are not attached to the spindle generate a checkpoint signal that prevents the start of anaphase until all the chromosomes are successfully attached and aligned on the equator of the cell.

Mitosis and cell specialisation

Key concepts you must understand

Chromosomes are made of a long molecule of DNA wound around proteins, which help with DNA packing. DNA is the important molecule as it codes for all the features of an organism. When you look at a dividing cell through a microscope, you can see chromosomes if they have been stained. Sometimes it is possible to see separate chromosomes, especially at metaphase during mitosis.

The daughter cells produced by mitosis have the same genetic information as the parent cell because all the cells have the same DNA. They are *genetically* identical. This makes it possible for cells to function together as one unit even if the daughter cells *express* different genes and function differently.

Key facts you must know

The stages of mitosis in an animal cell are outlined in Figure 45.

During interphase of the cell cycle DNA replication occurs so that each chromosome has two identical DNA molecules. These are wound around proteins to form chromatids that are joined together at the centromere. The two copies of each molecule of DNA within one chromosome are sister chromatids.

At the start of anaphase the DNA at the centromere separates so that the sister chromatids come apart. During anaphase, the sister chromatids move to opposite

Exam tip

Cytokinesis does not always follow mitosis. If there is no cytokinesis, a cell will have more than one nucleus. This is quite common, liver cells often have two nuclei. Striated muscle tissue, which makes up a large proportion of the human body, is made of multinucleate muscle fibres.

A **chromosome** is a thread-like structure made of DNA and protein; each chromosome becomes double-stranded after the DNA is replicated.

Mitosis is a type of nuclear division that gives rise to two daughter nuclei that are genetically identical to each other and to the parent nucleus.

A **chromatid** is one of the two 'strands' that make up a double-stranded chromosome.

Sister chromatids are genetically identical as they are the result of DNA replication and are joined at the centromere.

poles of the cell. They are pulled by the microtubules that make up the spindle. At telophase they form two nuclei with identical genetic information and the same number of chromosomes as the original cell.

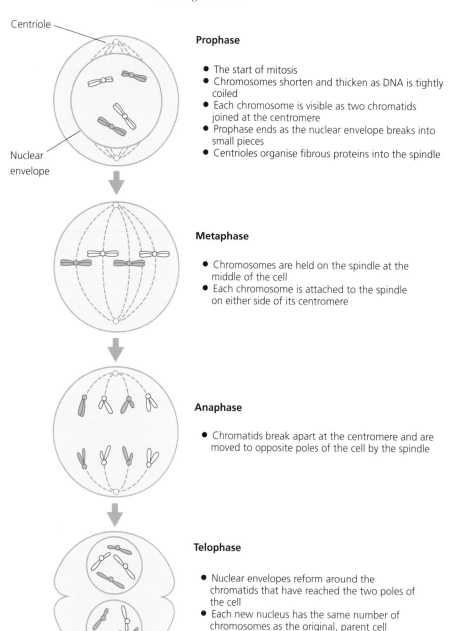

Prophase

- The start of mitosis
- Chromosomes shorten and thicken as DNA is tightly coiled
- Each chromosome is visible as two chromatids joined at the centromere
- Prophase ends as the nuclear envelope breaks into small pieces
- Centrioles organise fibrous proteins into the spindle

Metaphase

- Chromosomes are held on the spindle at the middle of the cell
- Each chromosome is attached to the spindle on either side of its centromere

Anaphase

- Chromatids break apart at the centromere and are moved to opposite poles of the cell by the spindle

Telophase

- Nuclear envelopes reform around the chromatids that have reached the two poles of the cell
- Each new nucleus has the same number of chromosomes as the original, parent cell
- The nuclei are genetically identical to each other

Figure 45 Stages in mitosis in an animal cell

Links

Replication of DNA must occur before mitosis. Each chromosome must have two DNA molecules so that they can be divided between the two new cells. The best way to understand the events that occur in mitosis is to watch time-lapse films of cells dividing on YouTube and some animations, for example on John Kyrk's website.

In multicellular organisms, the roles of mitosis are:

- growth
- repair following wounding or other damage
- replacement of cells and tissues
- asexual reproduction

Animals repair themselves following wounding or natural processes such as giving birth. Plants also repair themselves after wounding, for example when they are damaged by storms or eaten by grazers. New cells are produced in much the same way as during growth. Cells become worn out and need replacing, such as those lining the gut, which in humans are replaced each day.

Some animals and plants reproduce asexually, by budding or growing parts that separate from the parent. Asexual reproduction is really just a form of growth, so it also involves mitosis.

Yeast reproduces asexually by budding. When a yeast cell reaches a certain size part of the cell bulges out to form a bud. The nucleus divides in a way similar to that shown in Figure 45, but the nuclear envelope remains intact during the process. After the nucleus has divided, the bud may separate from the parent cell.

New individuals produced by asexual reproduction are genetically identical to the parent and are likely to survive in the same environment as the parent. This is especially the case with plants that spread by asexual reproduction, such as bluebells growing in a wood.

Cell specialisation in mammals

Figure 44 suggests that cells continue dividing all through their lives. Some cells remain able to divide throughout life — in mammals these are tissue stem cells that produce new cells to replace those that are worn out or die. Stem cells in bone marrow produce new red and white blood cells throughout life. Cells destined to become red blood cells (erythrocytes) go through a series of changes:

- ribosomes make many molecules of haemoglobin
- mitochondria and other organelles disintegrate
- the nucleus is extruded

These changes show how these unspecialised cells **differentiate** into mature cells that are able to carry out their specific function(s) for the body. Some white blood cells search out and engulf bacteria. These are known as neutrophils and they differentiate in bone marrow:

- Ribosomes make enzymes, which are packaged by the Golgi apparatus into lysosomes.
- The nucleus changes shape to become lobed — this allows neutrophils to move through the walls of blood capillaries.

Exam tip

You can make a large poster of the stages of mitosis using string, wool or modelling pipe cleaners for the chromosomes. Or make your own computer-generated images using illustration software.

In plants, meristematic cells in the cambium produce the cells of xylem and phloem (see Module 3 in Student Guide 2). In many trees and shrubs, the cambium produces new xylem during the growing season to give the familiar growth rings that you see in branches and tree trunks when they are cut in section.

Tissues, organs and organ systems

Key concepts you must understand

Biologists often talk about **levels of organisation**. Organelles carry out different functions in cells. This is the cell level of organisation. At the tissue level, similar cells cooperate to perform one or several functions. In organs, different tissues work together to perform a variety of major functions. In an organ system, organs work together to carry out major functions for the body, such as digestion, excretion and breathing.

Key facts you must know

Multicellular animals and plants are made up of large numbers of cells. Tissues are made of many cells that perform one or several functions. Often the cells are all of the same type. For example, epithelia are sheets of cells that line organs in the body and separate internal tissues from air, blood, food or waste that travel through tubes in the body. The outer part of your skin is an epithelium made of several layers of cells. Table 16 shows how some cells of mammals and flowering plants are specialised to perform their specific functions.

Table 16 Some specialised cells in mammals and flowering plants

Cell type	Function	Structural feature(s)
Mammals		
Erythrocyte (red blood cell)	Transports oxygen and carbon dioxide	No nucleus, to give more space for haemoglobin (see Module 3 in Student Guide 2)
Neutrophil	Engulfs bacteria by phagocytosis	Large number of lysosomes full of hydrolytic enzymes (see Module 4 in Student Guide 2)
Ciliated epithelial cell	Cilia beat back and forth to move fluids, for example mucus	Free end of cell is covered in a large number of cilia (see Module 4 in Student Guide 2)
Squamous epithelial cell	Forms a thin lining and provides a surface for diffusion	Thin flanges of cytoplasm around the nucleus to reduce distance for diffusion (see Module 3 in Student Guide 2)
Sperm cell	Transports chromosomes from male to the egg	Acrosome (modified lysosome); highly condensed chromosomes; flagellum
Flowering plants		
Palisade mesophyll cell	Photosynthesis	Large number of chloroplasts
Guard cell	Controls opening and closing of stomata	Unequal thickening of cell wall
Root hair cell	Absorption of water and ions from the soil	Long thin extension to give a large surface area (see Module 3 in Student Guide 2)

Cells forming squamous epithelia are flat and very thin. Looked at from above, each cell resembles a fried egg with the nucleus projecting like a yolk. Single layers of squamous epithelia, such as those lining the alveoli, are thin to help diffusion. Ciliated epithelial cells, such as those lining the airways in the lungs, have many cilia to move a fluid such as mucus over the surface.

Sperm cells are male gamete cells that are specialised for the delivery of chromosomes to the egg at fertilisation. They have:

- a streamlined shape
- chromosomes that are highly condensed in the head of the sperm
- an acrosome at the tip that contains hydrolytic enzymes to digest a pathway to the egg cells
- a flagellum to provide propulsion and mitochondria

Sperm cells are unusual in being specialised cells that do not form a tissue. Most specialised cells are joined together to form tissues. Many tissues contain several different types of specialised cell; others are made simply from one specialised cell, as shown in Table 17.

> **Knowledge check 27**
>
> Explain why ciliated epithelial cells contain many mitochondria.

Table 17 Some tissues in mammals and flowering plants

Tissue	Function(s)	Specialised cell(s)
Mammals		
Blood	Transport of gases, nutrients	Erythrocytes, neutrophils, monocytes, platelets
Ciliated epithelium	Movement of fluids, for example mucus in the airways	Ciliated epithelial cells (often with goblet cells)
Squamous epithelium	Lining alveoli and blood vessels	Squamous epithelial cells
Cartilage	Provides support and protection	Chondrocytes that secrete an extracellular matrix (see Module 3 in Student Guide 2)
Skeletal muscle	Contraction to move the body, for example intercostal muscles move the ribcage in breathing	Striated muscle fibres (multinucleate)
Smooth muscle	Contraction to move substances along tubes by peristalsis, for example in the gut, ureter and fallopian tubes	Smooth (unstriated) muscle cells (see Module 3 in Student Guide 2)
Cardiac muscle	Contraction to pump blood	Cardiac muscle cells (see Module 3 in Student Guide 2)
Flowering plants		
Mesophyll	Photosynthesis	Palisade mesophyll cells
Xylem	Transport of water and ions	Vessel elements organised into xylem vessels
Phloem	Transport of assimilates, for example sucrose and amino acids	Sieve tube elements organised into sieve tubes

Plant tissues

Xylem and phloem are the transport tissues of plants. Both are composed of three types of cell:

- cells that form tubes to provide a transport pathway
- parenchyma cells for storage and to provide energy
- fibres to help provide support

The transport cells in xylem are **vessel elements**. As these develop, they gain a strong, thickened cell wall and lose their cytoplasm, so becoming rigid and empty. They also lose their end walls and their cytoplasm and nuclei, so they form a continuous column of cells, known as a **xylem vessel**, which has little resistance to the flow of water.

The transport cells in phloem are **sieve tube elements**. These do not become thick walled and they keep some of their cytoplasm. The end walls are perforated to form sieve plates. Sieve tube elements form continuous columns known as **sieve tubes** for the transport of soluble substances, such as sucrose and amino acids, throughout the plant.

Organs

The human body has a number of different organs, such as heart, lungs, stomach, pancreas, spleen, brain, kidneys and liver. Examples of plant organs are leaf, stem and root. Organs are structures made of several tissues that work together to carry out a number of functions. For example, the leaf contains:

- epidermis for protection
- parenchyma for photosynthesis and storage
- xylem for transport of water
- phloem for transport of sucrose

Organ systems

Organs work together to carry out certain functions for animals. In the digestive system of a mammal, the mouth, oesophagus, stomach, small and large intestines, liver, gall bladder and pancreas work together to digest and absorb food and eliminate all the undigested material.

Chromosomes, life cycles and meiosis

Key concepts you must understand

Diploid cells have two sets of chromosomes. This means that there are two chromosomes of each type. In humans, the diploid number is 46; there are 23 pairs of chromosomes. We inherit our chromosomes from our parents. One set of chromosomes is inherited from our father, one from our mother. So one chromosome of each type is paternal and the other is maternal in origin.

Think about the sex chromosomes, X and Y, in boys. Males have one X and one Y. A boy inherits his X chromosome from his mother and his Y chromosome from his father. In the same way, he inherits one of each pair of chromosomes from his father and the other from his mother. Each pair of chromosomes is known as a **homologous pair**.

Key facts you must know

Homologous chromosomes have the same:

- shape and size
- position of the centromere
- genes

A **diploid** organism, cell or nucleus has two sets of chromosomes.

Homologous chromosomes carry the same sequence of genes; they are the same size and have their centromeres in the same relative position; they form **homologous pairs** (bivalents) during meiosis (see below).

It is possible to see homologous pairs of chromosomes under the microscope. Computers can scan photographs of cells taken at metaphase of mitosis and search for similarities, particularly in the banding patterns that chromosomes have when treated with certain dyes. The images of the chromosomes are then put into pairs, as shown in Figure 46.

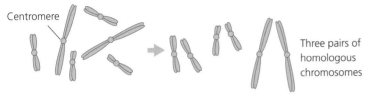

Figure 46 Homologous chromosomes

During sexual reproduction, gametes (sex cells) fuse at fertilisation. From generation to generation, the diploid number remains constant. There is no doubling of the chromosome number with each generation, as this would lead to cells with huge numbers of chromosomes and very large quantities of DNA. The diploid number stays constant from generation to generation because the number of chromosomes in gametes is half the diploid number.

Meiosis

Meiosis is the type of nuclear division that halves the chromosome number from diploid to **haploid**. It is sometimes called a reduction division because of this. In halving the chromosome number the two members of each homologous pair separate so that each new nucleus has one copy of each type of chromosome. Each haploid nucleus therefore has one set of chromosomes.

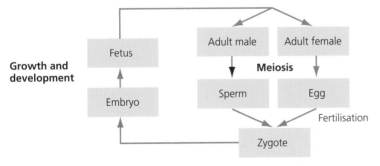

Figure 47 The position of meiosis in the human life cycle

The other role of meiosis is the generation of variation by:

- exchanging DNA between **non-sister chromatids** of homologous pairs during **crossing over**
- the random arrangements of paternal and maternal chromosomes at the equator of the cell, which is known as random or **independent assortment**

Meiosis is a type of nuclear division that gives rise to four haploid nuclei that are genetically different from each other and from the parent nucleus.

A **haploid** organism, cell or nucleus has one set of chromosomes.

Non-sister chromatids are chromatids of homologous chromosomes that are usually not identical.

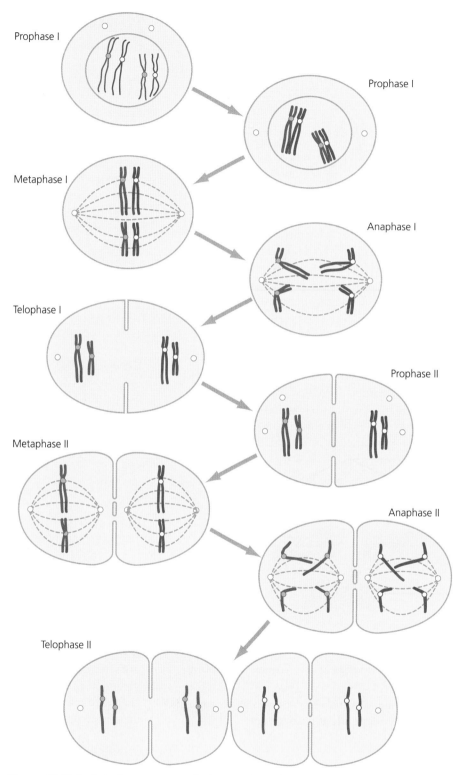

Figure 48 Meiosis to form sperm cells in an animal that has a diploid number of 4; maternal chromosomes are blue, paternal chromosomes are red

Exam tip

Questions on meiosis often involve sequencing the stages. To prepare for these make some drawings of the stages of meiosis and find some photomicrographs. Cut them up and put them in the correct sequence, using Figure 48 to help you. Then repeat without using the figure.

Prophase I

- Chromosomes condense so that they become shorter and thicker, and visible in the light microscope.
- Homologous chromosomes pair to form bivalents (maternal chromosomes are blue, paternal are red).
- Chiasmata (singular: chiasma) form to hold chromosomes together; non-sister chromatids join, break and exchange parts in crossing over.
- The nuclear membrane breaks up into small sacs of membrane, which become part of the endoplasmic reticulum; the centrioles replicate and move to opposite poles and form the spindle.

Metaphase I

- Bivalents move to the equatorial (or metaphase) plate across the centre of the cell.
- Paternal and maternal chromosomes in each bivalent position themselves independently of the others.
- Microtubules attach to the centromere of each chromosome.

Anaphase I

- Chromosomes (each with two chromatids) are pulled towards the poles by the shortening of the microtubules.

Telophase I

- Chromosomes reach opposite poles.
- Nuclear membranes reform to make two daughter nuclei that have half the number of chromosomes of the parent cell — these nuclei are **haploid**.
- Cytokinesis occurs — the cell surface membrane 'pinches in', leaving small cytoplasmic bridges between the cells.
- An interphase may occur between the divisions of meiosis I and II, in which case the chromosomes uncoil.

Prophase II

- Centrioles replicate and move to poles that are at right angles to those in meiosis I.
- Nuclear membranes break up.

Metaphase II

- Individual chromosomes align on the equator with their chromatids randomly arranged (important if crossing over has occurred in meiosis I).
- Microtubules attach to the centromeres.

Anaphase II

- Sister chromatids break apart at the centromere and move to opposite poles.

Telophase II

- Nuclear membranes reform.
- Cells divide to give four haploid cells that are genetically different from one another and from the parent cell.
- The haploid cells produced differentiate into sperm cells.

Exam tip

During prophase and metaphase of meiosis I chromosomes are 'double-stranded', as they have two chromatids joined together at the centromere. These are produced during interphase by replication. During anaphase II the chromatids separate to form two 'single-stranded' chromosomes.

You should look at microscope slides of pollen formation in the lily, *Lilium*. You can also find colour photographs of these stages online by searching for images of '*Lilium* microsporogenesis'. You should then compare slides and photographs with diagrams of meiosis and make sure that you can recognise the stages. Also look at animations of meiosis, for example on John Kyrk's website.

Crossing over

When they pair together homologous chromosomes attach at points known as chiasmata. The non-sister chromatids break at each **chiasma** and exchange parts, as you can see in Figure 49. This exchange of genetic material between non-sister chromatids gives chromosomes that are a mixture of maternal and paternal DNA.

Pairing of bivalent in early prophase I

Chiasma forms between non-sister chromatids in prophase I

Breakage and exchange of parts of non-sister chromatids

Figure 49 Crossing over between non-sister chromatids during prophase I

The letters **A**/**a** and **B**/**b** in Figure 49 represent two of the genes carried on this pair of homologous chromosomes. As a result of **crossing over** two of the chromatids have a new arrangement of the alleles of these two genes. This results in variation if they are inherited by the next generation.

Independent assortment

Independent (or random) **assortment** is the arrangement of homologous pairs of chromosomes on the equatorial plate of a cell during metaphase I. Figure 50 shows the two ways in which two homologous pairs can be arranged during metaphase in cells A and B.

Independent assortment generates variation because the haploid cells (gametes or spores) have a mixture of maternal and paternal chromosomes from the parent cell. Figure 50 shows random assortment with two pairs of chromosomes. Consider all the different arrangements that are possible with 23 pairs of homologous chromosomes.

Exam tip

Crossing over and independent assortment generate **genetic variation**. If you take the full A-level course, then you will appreciate the consequences of this when studying Module 6.

A **chiasma** (plural: chiasmata) is a point of attachment between chromatids during meiosis I to hold homologous chromosomes together.

Crossing over involves breakage and exchange of DNA between non-sister chromatids of an homologous pair.

Independent assortment is the random arrangement of pairs of chromosomes at metaphase of meiosis.

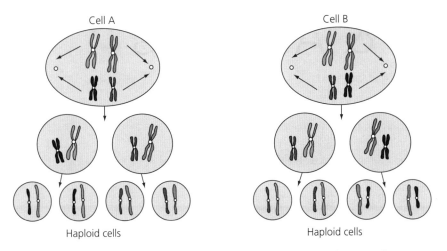

Figure 50 A and B show the two arrangements of maternal and paternal chromosomes in metaphase I responsible for independent assortment during meiosis in the formation of gametes

Summary

- The cell cycle comprises the changes that occur within a cell between one division and the next. About 90% of the cell cycle is spent in interphase when growth and the production of biochemicals and organelles occur. During the S phase DNA is replicated and proofread.
- Mitosis is a type of nuclear division. It takes up about 10% of the cell cycle and involves the movement of chromosomes; break-up and reformation of the nuclear envelope; and the production of the spindle, made from microtubules, that moves the chromosomes.
- Chromosomes that are the same size and shape and have the same genes are described as homologous. A diploid organism has a certain number of pairs of homologous chromosomes. Humans have 23, although in human males only part of the Y chromosome is homologous to the X.
- Mitosis maintains genetic stability as the two new nuclei have the same number of chromosomes as the parent cell and the genetic information is identical.
- Genetic stability is important in asexual reproduction, growth, replacement and repair.

- A stem cell is a cell that retains the ability to divide by mitosis to produce cells that develop into specialised cells. The process of changing into a specialised cell is differentiation.
- A tissue is a group of similar cells that perform the same function; an organ is a group of tissues that perform a variety of similar functions; an organ system involves several organs working together to perform a major body function. The survival of an organism depends on the successful cooperation of the different levels of organisation to carry out all the body functions.
- Meiosis is the type of nuclear division in which the chromosome number is halved. Each new nucleus is haploid as it has half the number of chromosomes of the parent cell. This is important in maintaining the diploid number from generation to generation in life cycles.
- Meiosis generates variation by crossing over between non-sister chromatids during prophase I and the independent assortment of different homologous pairs during metaphase I.

Questions & Answers

Exam format

At AS there are two exam papers. Questions in these two papers will be set on any of the topics from Modules 2–4 in the specification. In addition, there will be questions that will test your knowledge and understanding of practical skills from Module 1 and your ability to apply mathematical skills.

If you are taking AS biology, your exams will be as follows:

Paper number	1	2
Paper name	Breadth in biology	Depth in biology
Length of time	1 hour 30 minutes	1 hour 30 minutes
Total marks	70	70
Types of question	20 multiple-choice questions (1 mark each) and structured questions for 50 marks	Structured questions for 70 marks
Synoptic questions	Yes	Yes

At A-level there are three exam papers. Questions in these three papers will be set on any of the topics from Modules 2 to 6 in the specification. In addition, there will be questions that will test your knowledge and understanding of practical skills from Module 1 and your ability to apply mathematical skills.

If you are taking A-level, your exams will be as follows:

Paper number	1	2	3
Paper name	Biological processes	Biological diversity	Unified biology
Length of time	2 hours 15 minutes	2 hours 15 minutes	1 hour 30 minutes
Total marks	100	100	70
Section A	15 multiple-choice questions (1 mark each) and structured questions for 85 marks	15 multiple choice questions (1 mark each) and structured questions for 85 marks	Structured questions
Synoptic questions	Yes	Yes	The whole paper is synoptic

About this section

This section contains questions similar in style to those you can expect to find in your exam papers. The limited number of questions in this guide means that it is impossible to cover all the topics and all the question styles, but they should give you an indication of what you can expect.

The questions that follow are all based entirely on topics in this Student Guide for Module 2. The actual examination papers will not be like this — both papers at AS and all three papers at A-level will cover topics from Modules 2–4.

The AS-style questions start on this page and continue to p. 88. The A-level-style questions are on pp. 88–100. If you are not taking the AS papers you can still use the questions for your exam preparation.

The AS-style questions are similar to those you can expect in AS papers 1 and 2. The A-level-style questions are similar to those in A-level papers 1 and 2. Questions in A-level paper 3 are likely to cover topics and skills from all the modules and you have not covered enough at this stage to show you what questions from that paper will be like. Some of the A-level-style questions are set in the context of topics from Modules 5–6, to give you an idea of how you will be expected to use your knowledge of Module 2 in the A-level papers.

The answers to the 10 multiple-choice questions are on p. 100.

As you read through the answers to the AS-style questions, you will find answers from two students. Student A gains full marks for all the questions. This is so that you can see what high-grade answers look like. Student B makes a lot of mistakes — often these are ones that examiners encounter frequently. I will tell you how many marks student B gets for each question. The A-level-style questions only have model answers, similar to those of student A.

Comments

Each question is followed by a brief analysis of what to watch out for when answering the question (icon **e**). Some student responses are then followed by comments. These are preceded by the icon **e** and indicate where credit is due. In the weaker answers, they also point out areas for improvement, specific problems and common errors, such as lack of clarity, weak or non-existent development, irrelevance, misinterpretation of the question and mistaken meanings of terms.

■AS-style questions

Multiple-choice questions

Question 1

Four different plant cells were stained with iodine solution. All of the cells stained yellow except for one that contained many structures about 30 μm in length that stained black. The structures are:

A chloroplasts

B mitochondria

C starch grains

D vacuoles

(1 mark)

Question 2

A student made a drawing of a palisade mesophyll cell from a leaf. The length of the drawing was 65 mm and the actual length was 40 μm. Which is the magnification of the drawing?

A ×162.5

B ×1625

C ×2600

D ×16 250

(1 mark)

Question 3

Chloride ions are necessary for the proper functioning of salivary amylase.
Which describes the role of chloride ions?

A   coenzyme   **C**   prosthetic group

B   cofactor   **D**   substrate   (1 mark)

Question 4

Which row shows the correct monomers for the four carbohydrate molecules shown?   (1 mark)

	Cellulose	Glycogen	Starch	Sucrose
A	α-glucose	α-glucose	β-glucose	Glucose
B	α-glucose	β-glucose	α-glucose	Glucose + galactose
C	β-glucose	α-glucose	α-glucose	Glucose + fructose
D	β-glucose	β-glucose	β-glucose	Fructose

Question 5

Which process does not utilise metabolic energy from cells?

A   active transport   **C**   facilitated diffusion

B   exocytosis   **D**   phagocytosis   (1 mark)

ⓔ The answers to these multiple-choice questions are on p. 100, but try them first and then check your answers.

Structured questions

Question 6

For each of the statements in the table below, identify whether the description applies to:

■ only haemoglobin   ■ both haemoglobin and DNA
■ only DNA   ■ neither of them

You may use each response once, more than once, or not at all.   (5 marks)

Highly branched macromolecule	
Composed of amino acid monomers	
Contains nitrogen atoms	
Composed of many monomers linked by phosphodiester bonds	
Contains phosphate	

ⓔ When you answer questions like these make sure that you follow the rubric — for example, in this case the instructions tell you that you do not have to use all of the responses.

Student A

Highly branched macromolecule	Neither of them
Composed of amino acid monomers	Haemoglobin
Contains nitrogen atoms	Both of them
Composed of many monomers linked by phosphodiester bonds	DNA
Contains phosphate	DNA

ⓔ **5/5 marks awarded**

Student B

Highly branched macromolecule	Hemoglobin
Composed of amino acid monomers	DNA
Contains nitrogen atoms	Hemoglobin
Composed of many monomers linked by phosphodiester bonds	DNA
Contains phosphate	Neither of them

ⓔ **1/5 marks awarded** Student B has used the American spelling for haemoglobin but the meaning is clear, although haemoglobin is not the correct answer in each case. A similarity between DNA and proteins is that they are not branched. It is a common error to state that DNA is made up of amino acids; the monomers for nucleic acids are nucleotides (see p. 30). DNA is the correct answer for row 4, but proteins do not contain phosphate. None of the amino acids has phosphate as part of its R group.

Question 7

The table shows some features of cells. Complete the table by:

■ putting a tick (✓) or a cross (✗) to indicate whether the feature is present in prokaryotic cells and eukaryotic cells or not

■ writing a brief statement to describe a function of each feature   (6 marks)

Feature	Prokaryotic cell	Eukaryotic cell	Function
Plasma membrane			Controls movement of substances in and out of cells
30nm ribosomes			
Mitochondrion			
Pili			
Capsule			Provides protection
Chloroplast			

ⓔ Always take care over putting in ticks and crosses — it might look an easy type of question, but often requires careful thought.

Student A

Feature	Prokaryotic cell	Eukaryotic cell	Function
Plasma membrane	✓	✓	Controls movement of substances in and out of cells
30 nm ribosomes	✗	✓	Synthesise proteins
Mitochondrion	✗	✓	Aerobic respiration
Pili	✓	✗	Allow movement of DNA from cell to cell
Capsule	✓	✗	Provides protection
Chloroplast	✗	✓	Photosynthesis

ⓔ 6/6 marks awarded

Student B

Feature	Prokaryotic cell	Eukaryotic cell	Function
Plasma membrane	✓	✓	Controls movement of substances in and out of cells
30 nm ribosomes	✗	✓	Make proteins
Mitochondrion	✗	✓	Respiration
Pili	✓	✗	Holds two bacteria together
Capsule	✗	✗	Provides protection
Chloroplast	✗	✓	Absorbs light energy

ⓔ 4/6 marks awarded Student B has given the correct responses for the eukaryotic cell, but it is not clear what is intended for the answer for capsule in the prokaryotic cell. This could be a tick that has been converted into a cross or a cross that has been changed to a tick. In cases like this, always rewrite your answer in a blank space and then cross out your original answer. Do not try to change it as student B has done.

Student B's other answers are correct, except for the function of the mitochondrion. Respiration is a series of chemical reactions some of which occur in the cytoplasm surrounding the organelles and some of which occur in mitochondria. It is the series of reactions that occur when oxygen is available that occur in mitochondria and hence the answer is *aerobic* respiration.

Question 8

The diagram shows the replication of a small part of a molecule of DNA.

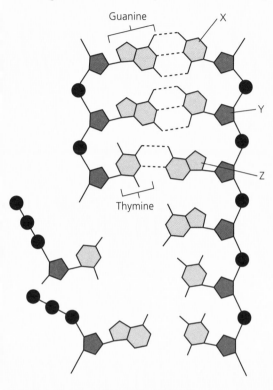

(a) (i) Name X, Y and Z in the diagram. (3 marks)

ℯ When naming parts of DNA or RNA take care over spelling. Examiners may insist on spelling being correct in cases where misspellings could indicate confusion with another compound. For example, the base thymine is often misspelt as thiamine, which is a vitamin.

(ii) Using information from the diagram, explain why the two polynucleotide strands in DNA are described as anti-parallel. (2 marks)

ℯ Questions that start like this often give you some help in phrasing your answer. Study the diagram carefully for ideas to use in the answer.

(b) Describe how DNA is replicated. You may refer to the diagram in your answer. (6 marks)

ℯ Look carefully at the diagram and use the information about the polynucleotides, their bases and the base pairing to make the new molecule of DNA alongside the template polynucleotide that is visible on the right of the diagram.

(c) The table shows the relative quantities of purine bases and pyrimidine bases as determined in three samples of DNA.

Source of DNA	Purines		Pyrimidines	
	Adenine	Guanine	Cytosine	Thymine
Human liver cells	30.3	19.5	19.9	30.3
Human sperm	27.8	22.2	22.6	27.5
Yeast	31.7	18.3	17.4	32.6

Comment on the figures in the table. (4 marks)

ⓔ Study the information in tables very carefully. Read the information that comes before the table to make sure you understand what the data are about. Then look carefully at the row and column headings and any units. Then look across each row and down each column looking for trends or patterns. Circle any pairs of figures that you could use for data quotes. Recall the subject knowledge that you have about the topic. Here you should recognise base pairs.

Student A

(a) (i) X — cytosine

Y — deoxyribose

Z — adenine

ⓔ **3/3 marks awarded**

Student B

(a) (i) X — cysteine

Y — sugar

Z — adenosine

ⓔ **0/3 marks** Student B has made three errors here:

- confusing cysteine (which is an amino acid) with cytosine
- not being specific about the name of the pentose sugar in DNA
- misspelling adenine as adenosine

The 'backbone' of DNA is sometimes referred to as the sugar–phosphate backbone and you can use this expression when describing replication, for example. But here you should always be as specific as you can with the names you give. 'Sugar' could be a hexose (e.g. glucose), a pentose (e.g. ribose) or a disaccharide (e.g. sucrose). Pentose would be a slightly better answer, but that includes ribose, which is found in RNA and not DNA, so cannot be correct either.

Student A

(ii) Deoxyribose in the diagram is organised so that carbon 3 points upwards and carbon 5 points downward in the polynucleotide on the right, as here:

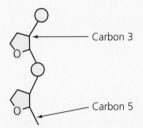

Carbon 3

Carbon 5

On the opposite polynucleotide sugar-phosphate backbone, the deoxyribose molecules are pointing in the opposite direction, with carbon 3 at the bottom and carbon 5 at the top.

ⓔ 2/2 marks awarded Student A has given a detailed answer, realising that it is the orientation of the deoxyribose residues in the sugar–phosphate backbone that is the key point to make. The carbon atoms are numbered 1–5.

Student B

(ii) The polynucleotides point in opposite directions.

ⓔ 0/2 marks awarded The detail expected is not here. Make sure that you can define all the terms given in the learning outcomes of the OCR specification. Some of them are shown as key terms in the margin of the Content Guidance section of these Student Guides.

Student A

(b) The two polynucleotides separate as the hydrogen bonds between the bases are broken. Each polynucleotide acts as a template strand. DNA polymerase moves along each strand. As it moves along, free nucleoside triphosphates (dNTPs or 'energised' nucleotides) are assembled against the template strands. In the example, a dNTP with a T is about to be paired with an A and a dNTP with a C has just been paired with a G at the top of the diagram.

Two of the phosphates split away from the dNTPs, providing energy for the formation of a phosphodiester bond between sugar (deoxyribose) and phosphate. This is semi-conservative replication, as each new DNA molecule has the template strand and a new strand that has been made alongside it.

ⓔ 6/6 marks awarded Student A has easily gained full marks for this question and has made good use of the material in the diagram.

Student B

(b) The strands separate and one strand acts as a template for making another new strand. Bases are matched together by base pairing and these are joined together to make a new strand.

ⓔ 0/6 marks awarded Student B has not stated that both polynucleotides act as templates, has not used the information in the diagram and has not used appropriate terminology. No marks are awarded even though base pairing has been mentioned — the examiner would expect information from the diagram to be used before awarding a mark.

Student A

(c) In liver cells the relative quantities of adenine (A) and thymine (T) are identical. In the other samples they are almost the same. The quantities of cytosine and guanine are different from A and T but the quantity of C is always almost the same as the quantity of G. This is explained by base pairing between purines and pyrimidines in DNA. The base pairs are A–T and C–G. The figures for sperm are slightly different from the liver cells because sperm are haploid and do not have DNA identical to the diploid liver cells.

ⓔ 4/4 marks awarded Student A has solved the problem. It is not just a coincidence that the relative quantities of A and T are the same or that the relative quantities of C and G are the same. If you think of the structure of DNA you will immediately realise that the data in Table 1 support the idea of base pairing between purines and pyrimidines.

Student B

(c) The numbers for adenine are all about 30, for guanine they are about 20, for cytosine they are about 20 (although yeast is lower for guanine) and for thymine about 30. These are the four bases found in DNA and the figures in the table show that humans and yeast have DNA with the same amounts of these four bases. This could be a coincidence or it could be that all DNA is like this. More data are needed to be sure.

ⓔ 0/4 marks awarded Student B's answer consists simply of a description of the data rather than anything insightful. The command word *comment* implies that you can describe, state, explain and evaluate. When presented with data that you have not seen before always study them carefully, look for any trends or patterns and *use your knowledge*.

Question 9

A student investigated the effect of increasing enzyme concentration on the activity of a protease. Milk protein was used as the substrate. Different concentrations of the enzyme were made up as shown in the table. $2\,cm^3$ of each protease solution was added to $10\,cm^3$ of the milk solution and the time taken for the cloudiness to disappear was recorded. The investigation was carried out at 35°C.

The rate of reaction was calculated as $(1/t) \times 1000$, where t = time taken for disappearance of the cloudiness. The student's results are shown in the table.

Concentration of protease/g dm⁻³	Time taken for cloudiness to disappear/s	Rate of reaction × 10³/s⁻¹
0.0	0	0.0
1.0	555	1.8
2.5	166	6.0
5.0	92	10.9
7.5	73	13.7
10.0	49	

(a) (i) Calculate the rate of reaction when the concentration of the enzyme was $10\,g\,dm^{-3}$. Show your working.

(2 marks)

ⓔ When you complete a table like this, make sure that you look at the figures that have already been calculated. This is a useful check to make sure that you use the right method.

(ii) Draw a graph of the results in the table.

(5 marks)

ⓔ You will be provided with a 2 mm grid on the examination paper to draw any graphs like this. There will be enough space to draw axes that will use more than half the space available. Use a pencil for all your writing on the graph; do not use a pen in case you make a mistake and need to erase something.

(iii) Use your graph to describe and explain the student's results.

(4 marks)

ⓔ Even if the question does not advise you to use the data, you should do this anyway when describing results.

(b) State *two* factors that would be kept constant and state how the student would keep them constant in this investigation.

(4 marks)

ⓔ Remember the four factors that influence enzyme activity. One of them is the independent variable, so you have a choice of two from three.

(c) Explain how the student could use a colorimeter to follow the course of each reaction to find the initial rate of reaction.

(5 marks)

ⓔ You should know about using a colorimeter from the Benedict's test (p. 28). This question is asking you to apply your knowledge to a new situation — there will be quite a few questions like this in the examination.

Student A

(a) (i) $\frac{1000}{49} = 20.4\,s^{-1}$

ⓔ 2/2 marks awarded

Student B

(a) (i) $20.4\,s^{-1}$

ⓔ 2/2 marks awarded Student B has not followed the instruction to show working, but examiners will often give full marks even if working is not shown. However, it is always safer to write down the calculation because you may make a mistake and the examiner can still give 1 mark here for correct working.

Note that both students have expressed the answer to one decimal place to match the other rates given in the table. They will have rounded down from 20.41 to 20.4. Often the examiner may ask you to put the answer in the table. Always look for this instruction and follow it.

Student A

(ii)

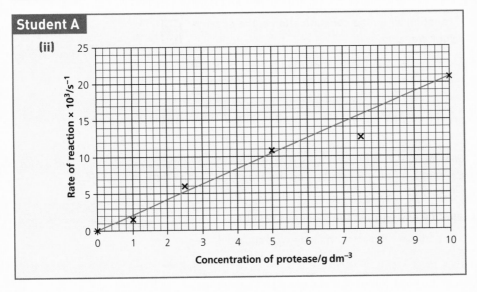

ⓔ 5/5 marks awarded Student A has correctly:

- put the independent variable on the *x*-axis
- selected suitable scales for the two variables
- labelled the axes with the correct names of the two variables
- included the units exactly as they are given in the table
- plotted points accurately

Student A has put in a line of best fit. This is perfectly acceptable so long as it starts at the origin (0, 0) and does not go beyond the last plotted point (here at $10\,g\,dm^{-3}$).

Student B

(ii)

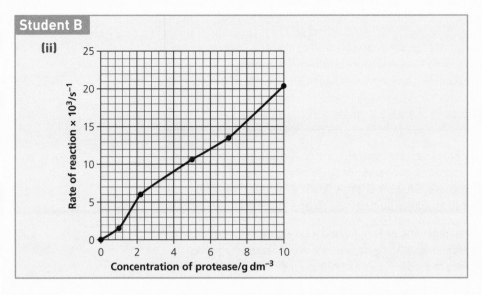

e 4/5 marks awarded Student B loses a mark for not showing the data points clearly and using a blunt pencil for drawing the line. The points are joined with straight lines, which is also acceptable, although it is quite clear here that the rate of reaction is directly proportional to the concentration of protease and a straight line would indicate the relationship between rate and enzyme concentration. They could argue that there were no replicates (repeat readings) for each concentration and not enough intermediate concentrations between $0\,g\,dm^{-3}$ and $10\,g\,dm^{-3}$ to be certain about where to put the line of best fit. These would be points worth making if there had been an evaluation question here.

Student A

(iii) The rate at which the milk protein is broken down increases as the concentration of the enzyme increases: the relationship is linear. The rate at $10.0\,g\,dm^{-3}$ ($20.4\,s^{-1}$) is double that at $5.0\,g\,dm^{-3}$ ($10.9\,s^{-1}$). The cloudiness disappears because the milk protein is hydrolysed by the protease breaking the peptide bonds between the amino acids. As the concentration of enzyme increases, there are more active sites available and there are more collisions between protease molecules and the milk protein molecules. More enzyme–substrate complexes can form.

e 4/4 marks awarded Student A has given a description of the results and illustrated it with a comparative data quote. The answer goes on to explain why the cloudiness has disappeared and why the rate increases as there is an increase in concentration of enzyme.

> **Student B**
>
> **(iii)** There is a direct relationship between rate and concentration of enzyme. When there is no enzyme, there can be no reaction (or the rate is so slow that it cannot be measured) as there are no active sites available. As more enzyme is available, there are more active sites for the substrate molecules, so the rate increases.

e **1/4 marks awarded** Student B has said that there is a 'direct relationship' without saying what that is. The answer has not been developed to explain the effect of an increase in active sites. Student B gains 1 mark for referring correctly to active sites but has not used any data, so has missed an easy mark here.

When you are describing data in graphs or tables, always include a data quote to illustrate what you are describing. Often a data quote must be comparative, for example the rate has increased from X to Y as the enzyme concentration was increased from A to B. Don't forget to use the units when giving data quotes.

> **Student A**
>
> **(b)** pH: the student would use a buffer solution at a set pH — perhaps the optimum pH for the enzyme.
>
> Temperature: the test tubes would be placed into a thermostatically controlled water bath at a set temperature — perhaps the enzyme's optimum temperature.

e **4/4 marks awarded**

> **Student B**
>
> **(b)** Concentration of enzyme: the amount of the enzyme added to each test tube must be the same.
>
> Concentration of milk: the amount of milk added to each test tube must be the same.

e **1/4 marks awarded** Student B shows some confusion. The first answer is the independent variable in this investigation and is changed each time rather than being kept the same throughout. No marks are awarded for this.

In the second answer there is confusion over the use of the word 'amount'. The concentration of the milk protein is a variable that must be controlled, so the student has 1 mark for stating the variable, but 'amount' here could mean concentration (mass of milk powder per unit volume of water) or it could mean the volume of milk solution added. Both of these options would be correct, but the student has not given a clear answer, so only gains 1 mark.

Student A

(c) 10 cm³ of the milk solution with 2 cm³ of distilled water could be put into a colorimeter and a reading taken for absorbance. This is equivalent to the very start of the reaction just as the enzyme is added (time = 0 seconds). The first enzyme solution (1 g dm⁻³) is added to the milk solution and immediately put into the colorimeter. The absorbance readings are taken every 10 seconds.

This is repeated with fresh milk solutions and the other enzyme concentrations, as in the table. This would be done at room temperature rather than 35°C, so the rate would be slower than in the table. A graph is drawn and the initial rate is calculated by taking the tangent, as shown here:

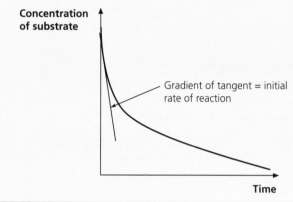

Concentration of substrate

Gradient of tangent = initial rate of reaction

Time

ⓔ 5/5 marks awarded

Student B

(c) When the milk and enzyme solution are mixed, the cloudiness disappears. It is difficult to tell exactly when the cloudiness has disappeared, so the student could put the test tube into a colorimeter and wait until the reading is 0. The student can then record the time taken to reach a reading of 0 in the colorimeter and this will be a better measurement of the rate of the reaction than just looking at it with your eyes. People will differ in the way in which they see the cloudiness and they may not always get the same result when looking at the same test tube.

ⓔ 0/5 marks awarded The examiner has not asked the students to improve the way in which the results of the student's investigation are taken — which is the question that student B has answered. To find the initial rate, readings are taken over a short period of time and plotted on a graph, as student A describes. Calculating 1000/t (t = time taken for milk solution to clear) as the rate of the reaction is fine, but it does not give you the rate at the very beginning when substrate concentration is not limiting.

Another way to use the colorimeter is to connect it to a data logger or computer. The software will plot a graph to show the decrease in absorbance with time. It is possible to put test tubes into some colorimeters. With others, the reaction mixtures are placed into cuvettes, which are transparent, plastic containers square in cross section.

Question 10

(a) Some molecules may cross plasma (cell surface) membranes by simple diffusion. Glucose, however, does not. Explain why glucose cannot cross membranes by simple diffusion. (2 marks)

(e) Sometimes you can be asked why certain things *do not* happen. This is one of those questions. Think about the properties of glucose that prevent it crossing membranes by simple diffusion. Think also about the ways in which substances cross membranes and choose the right one.

(b) A student carried out an investigation to find out whether cells take up glucose by facilitated diffusion, by placing animal cells into nine different solutions of glucose. The student determined the rate of uptake of glucose across the plasma membrane into the cells for each solution. The results are shown in the graph.

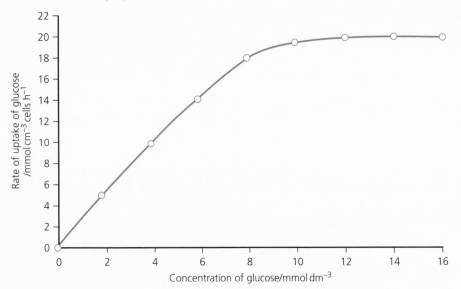

Using the information in the graph, explain how the results of the student's investigation support the idea that glucose enters cells by facilitated diffusion. (2 marks)

(e) To answer this question you must use the data from the graph. Look at the pattern of increasing uptake, with a plateau where the rate remains constant. The uptake does *not* stop between the concentrations of 12 and 16 mmol dm^{-3}!

(c) The student next investigated the movement of water into and out of cells by taking two pieces of epidermis from the scale leaf of a red onion bulb and putting one in distilled water (**P**) and the other into a concentrated solution of salt (**Q**). After several minutes, the two pieces of epidermis were mounted on microscope slides using the liquids in which they had been immersed. The student took the following photographs of cells from each slide.

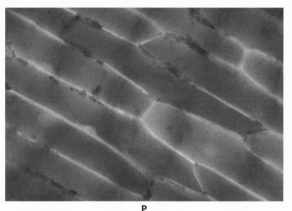
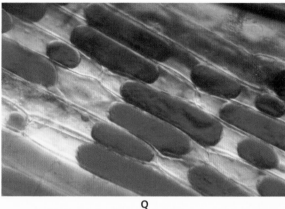

P   Q

Explain why the cells in **Q** have a different appearance compared with those in **P**. (4 marks)

e Start your answer by describing the appearance. There will probably be a mark for that even though the question says explain. You have to state the difference first before you can explain it.

(d) The student put a drop of some mammalian blood on a microscope slide and added a drop of a concentrated salt solution. The student then looked at the appearance of the red blood cells with the high power of a light microscope. Describe and explain the appearance of the red blood cells after they had been mixed with a concentrated salt solution. (4 marks)

e Note that this question says 'describe *and* explain' — it requires you to respond to both command words if you are going to get full marks. Before starting it is a good idea to write *description* on the first line and *explanation* on a line further down the answer space to remind you to cover both aspects.

(a) Glucose is a polar molecule. The membrane consists of a phospholipid bilayer. Phospholipid molecules each have two hydrophobic fatty acid tails. These tails face each other, forming a hydrophobic interior to the membrane that does not permit the passage of (large) hydrophilic molecules like glucose.

e **2/2 marks awarded**

(a) Glucose is a large molecule and cannot diffuse across the cell membrane because the gaps between the cell membrane are tiny.

e **1/2 marks awarded** Student B needs to refer to the polar nature of glucose, not its size. 'Gaps *between the cell membrane*' does not really make sense. The student may be thinking about channel proteins as pores that run through the membrane. To gain a second mark there should be something here about the phospholipid bilayer, as given by student A.

Student A

(b) For low concentrations of glucose outside the cell, glucose diffuses into the cells, moving down its concentration gradient (as glucose concentration is higher outside the cells than inside them). The rate of uptake increases with concentration. However, after a concentration of about $12\,mmol\,dm^{-3}$, the rate of glucose uptake becomes constant because limited numbers of carrier proteins specific to glucose are available. This shows that it is facilitated diffusion, not simple diffusion.

---

e **2/2 marks awarded**

Student B

(b) As the concentration of glucose increases, the rate of uptake of glucose also increases but then remains constant and does not decrease or increase.

---

e **1/2 marks awarded** The question asks for an *explanation* using evidence from the graph, which is what student A has done to good effect. Student B has simply *described* the graph, which gains 1 mark, but there is no explanation. If uptake was by simple diffusion you would expect the rate to continue to increase as the concentration of glucose in the external solution increased rather than reaching a plateau at high concentrations of glucose. The carrier proteins have become the limiting factor here — as student A states, there is a limited number of them and there is a limit to the rate at which glucose molecules can move through them.

Student A

(c) In Q, water has diffused by osmosis out of the cell down its water potential gradient. Most of the water has come from the vacuole and this has decreased in volume, pulling the cytoplasm away from the cell wall. The salt solution fills the space between the cell membrane and the cell wall. P has been in water, so there has been diffusion of water into the cells, making them fully turgid. The cell wall exerts a pressure potential, so the cells reach a maximum size and do not continue to expand.

---

e **4/4 marks awarded**

Student B

(c) The cell shows plasmolysis. There is a big space between the cell wall and the cell membrane because water has moved down a concentration gradient.

ⓔ 0/4 marks awarded Student B has not looked at the mark allocation for part (c). There are three ideas in the answer, so even if they were correct they would only gain 3 marks. For a 4-mark question there should be four ideas, if not five or six to be on the safe side.

Student B has described the appearance of the cell rather than explained what happened. Notice that student A's answer correctly includes the term *water potential*. You should never refer to a 'concentration gradient' for water since the movement of water is dependent on a number of factors, not just quantity of water. Sometimes examiners will instruct you to use the term *water potential* in your answer.

The introduction to part (c) states that the student was investigating the movement of water. This is a good clue that the question is about osmosis and water potentials.

Student A

(d) The cells will have a crinkly appearance, which is known as crenation. The concentrated salt solution has a lower water potential than blood plasma and it is lower than the cells. As a result, water has diffused out of the cells by osmosis down the water potential into the surrounding salt solution.

ⓔ 4/4 marks awarded

Student B

(d) The cells will explode because they will absorb the salt from the salt solution. This will give them a lower water potential than the surroundings so they absorb water as well and this causes them to burst as they have no cell wall.

ⓔ 0/4 marks awarded Student B has assumed that salt diffuses into the cells, lowers the water potential so water follows and they burst. Red blood cells do not absorb sodium or chloride ions that easily so the opposite happens, as described by student A. Red blood cells will burst if put into distilled water (with a water potential of 0 kPa) and this happens because they do not have a cell wall, unlike the onion cells in part (c).

Question 11

Mammalian cells maintained in culture can be made to fuse together. Cell fusion experiments have been used to investigate the roles of cell signalling substances produced inside cells to control the cell cycle.

In an experiment, mammalian cells at different stages of the cell cycle were fused together. The diagram shows what happens to the nuclei of stem cells at different stages of the cell cycle after they were fused together.

Experiment	1		2		3	
	S	G$_1$	S	G$_2$	M	G$_1$

Cells at different stages of the cell cycle

Cell fusion

Stage of the cell cycle shortly after fusion

(a) State the property of plasma membranes that allows two mammalian cells to fuse together. (1 mark)

ⓔ The command word *state* means that you should be able to recall the answer. This is something you should know.

(b) Cell fusion techniques were used to test the hypothesis that cells release substances to promote stages in the cell cycle at times when they are ready to progress from one stage to the next.
Use the diagram to discuss the evidence to support this hypothesis. (6 marks)

ⓔ This question requires an extended answer, which should be written in continuous prose and not in bullet points. Planning the answer is important, so look at the diagram carefully, answer separately for each of the three examples, and perhaps end with a general statement about control of the cell cycle as revealed by the data.

Student A

(a) The phospholipids in cell membranes are fluid, so in cell fusion the phospholipids of the two membranes flow together.

ⓔ 1/1 mark awarded

Student B

(a) Membranes are fluid mosaics, so when the cells fuse the membranes join together.

ⓔ 1/1 mark awarded Both students have used the term *fluid* in their answers, and this is the property that is the appropriate one. Student A gives a better answer as the fluidity is ascribed correctly to the phospholipids in membranes, but both students gain the mark.

Student A

(b) Experiment 1 — The nucleus in G_1 has entered the S phase so any signalling compounds to start DNA replication must be present and could have come from the nucleus or cytoplasm in the cell in stage S. The G_1 checkpoint has been passed.

Experiment 2 — The nucleus in G_2 has already been through the S phase so any signalling compound from the cell in the S phase has no effect. There is no point in replicating twice. The G_2 nucleus has not yet passed its checkpoint to start mitosis.

Experiment 3 — The metaphase nucleus has gone through division; the G_1 nucleus has also started to divide (perhaps by entering prophase) but has not gone any further. The G_1 nucleus would normally go through the S phase, so a signalling compound from the cell in metaphase might have stimulated it to start dividing even though it was not ready. As the S phase has not occurred it does not have chromosomes with two chromatids each so cannot divide.

ⓔ 6/6 marks awarded

Student B

(b) During the cell cycle there are three control points known as checkpoints. At each checkpoint the cell makes sure that everything is ready to continue to the next stage. These checkpoints are at the end of G_1, the end of the S phase and at the end of metaphase. In experiments 1 and 2, the nuclei have started to divide, but only that from G_2 has divided correctly. In experiment 3 the cell in metaphase carries on dividing, but the cell in G_1 has started dividing but has not got passed prophase.

ⓔ 1/6 marks awarded Student B has started the answer with a good outline of the control of the cell cycle, but has not spent enough time *answering the question*. Answers to this question have to find evidence from the figure to support the hypothesis, and they rely on observation and insight, not just the recalling of knowledge. Student B gets 1 mark for identifying that checkpoints are important, but has not applied that knowledge to the answer.

e Overall, student B scores 16 marks out of 65. This is not enough for an E grade. Marks were lost for a number of different reasons:

- Some answers are not developed fully (e.g. Q.8aii, Q.11b).
- An answer has been changed, but the amendment is not clear (e.g. Q.7).
- Command words have not been followed (e.g. Q.8c and 10b, where the student described rather than explained).
- Clues in the question have not been spotted (e.g. Q.8c).
- The mark allocation has not been followed (e.g. Q.10c).
- Data provided have not been used in answers (e.g. Q.9aiii).
- Answers are not precise enough (e.g. Q.9b, where the student uses the word 'amount' and Q.8ai, where 'sugar' is given instead of the name of the specific sugar.
- Common errors have been made (e.g. Q.6, where DNA was said to be made of amino acids).
- Not answering the question (e.g. Q.9c).
- Using a thick pencil to draw the lines on a graph (e.g. Q.9aii).
- Technical words are misspelt (e.g. Q.8ai).

■ A-level-style questions

Multiple-choice questions

Question 1

Two glucose solutions were tested with Benedict's solution. After they were cooled the solutions were filtered and the filtrate was tested with a colorimeter. The same procedure was repeated with distilled water instead of glucose solution. The results are shown in the table.

Glucose concentration/$g\,dm^{-3}$	Absorbance
0.0	2.00
5.0	1.84
100.0	0.00

Which is the explanation for the low reading for $100\,g\,dm^{-3}$ glucose solution?

A Copper oxide has been removed by the filtering.

B Benedict's solution has copper ions that absorb light.

C The high concentrations of glucose are very cloudy.

D Glucose has reduced all the copper ions in the Benedict's solution.   (1 mark)

Question 2

The graph shows the activity of an enzyme under five different conditions.

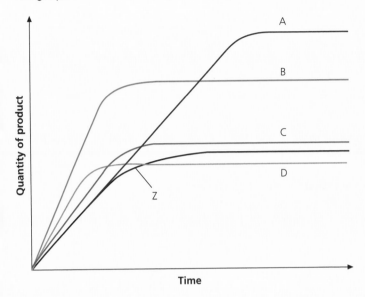

Z shows the quantity of product produced at intervals of time at 15°C. Which shows the effect of doubling the concentration of the substrate?

(1 mark)

Question 3

Which row shows the correct distribution of nucleic acids in a plant cell?

(1 mark)

	Nucleus	Cytoplasm surrounding organelles	Mitochondrion	Chloroplast
A	DNA and tRNA	DNA and mRNA	mRNA and tRNA	DNA and mRNA
B	mRNA and tRNA	mRNA, tRNA and rRNA	DNA and mRNA	mRNA and rRNA
C	DNA and mRNA	mRNA, tRNA and rRNA	DNA, mRNA, tRNA and rRNA	DNA, mRNA, tRNA and rRNA
D	mRNA and rRNA	DNA, tRNA and rRNA	DNA and rRNA	DNA and rRNA

(1 mark)

Question 4

The template (non-coding) strand of DNA has the following sequence of bases:
    ATGCCCCTCAGATTT

Which shows the anticodons for the first five tRNA molecules?

A ATG CCC CTC AGA TTT

B TAC GGG GAG TCT AAA

C UAC GGG GAG UCU AAA

D AUG CCC CUC AGA UUU

(1 mark)

Question 5

Which event occurs in meiosis but *not* in mitosis?

A   separation of chromatids

B   movement of centrioles to the poles

C   formation of chiasmata

D   break-up of nuclear membranes   (1 mark)

Structured questions

Question 6

Insulin is a protein hormone secreted by β cells in the pancreas. The diagram shows the interrelationships between the organelles involved in production and secretion of insulin from a pancreatic cell.

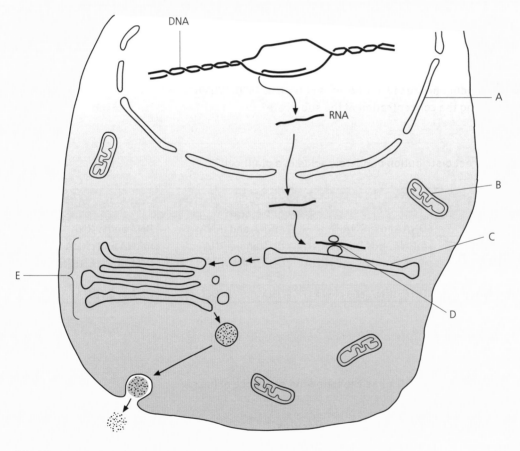

(a) (i)   Name the structures A, B and C.   (3 marks)

💬 When asked to name, you only have to give a very brief answer. You will usually be given a line on which to write each name, however brief.

(ii) Describe the functions of structures D and E in the production of protein in the pancreatic cell shown. (4 marks)

ⓔ You are not asked to name structures D and E and you are unlikely to gain any marks for naming them, so concentrate on describing their functions.

In type 1 diabetes, the β cells in the pancreas are destroyed by the body's immune system and do not secrete insulin. People with this form of diabetes use biosensors to monitor the concentration of glucose in their blood and have regular injections of insulin.

(b) Explain:
   (i) what is meant by a *biosensor* (2 marks)
   (ii) why insulin cannot be taken by mouth (2 marks)

Researchers have investigated the possibility of using stem cells as a treatment or even a cure for type-1 diabetes. Stem cells are removed from a person with diabetes, grown in the laboratory, stimulated to become insulin-secreting cells and implanted into the body. Trials have been most successful when a person's own stem cells are used.

(c) (i) What are stem cells? (3 marks)
   (ii) Explain the advantages of using stem cells for this treatment. (3 marks)
   (iii) Explain the reasons for using a person's own stem cells rather than those from a person who does not have diabetes. (3 marks)

ⓔ This question relies on knowledge of immunity from Module 4.

Student answers

(a) (i) A = nuclear envelope; B = mitochondria; C = rough endoplasmic reticulum

ⓔ **3/3 marks awarded** This answer identifies the three structures correctly. Although label B is only pointing to one mitochondrion, examiners will accept *mitochondria* in cases like this.

(ii) The ribosome (D) assembles amino acids to make polypeptides. This process is called translation. The polypeptides move through the ER to reach the Golgi apparatus (E), where they are modified and put into vesicles. The vesicles bud off from the Golgi and move to the cell surface.

ⓔ **4/4 marks awarded** Here there are two correct statements about *each* structure. Ribosomes assemble amino acids to make polypeptides and proteins are modified inside the Golgi apparatus and put into vesicles.

(b) (i) A biosensor uses biological molecules to measure the quantity of a particular substance, such as glucose. The biosensor that detects glucose has an enzyme that catalyses a reaction that leads to a small electric current. The current is detected by an electrode, amplified and displayed as a digital read-out.

Questions & Answers

ⓔ 2/2 marks awarded This gives a good outline of a biosensor as required by the specification. More detail might involve the name of the enzyme and the reaction that it catalyses.

> **(b) (ii)** Insulin is a protein and will be digested in the gut. Protease enzymes like trypsin will hydrolyse the peptide bonds in between the amino acids in insulin.

ⓔ 2/2 marks awarded This question tests your knowledge of insulin and the roles of proteases, such as trypsin. This is the obvious answer, although it is pepsin in the stomach that will start the digestion of any insulin taken by mouth. Trypsin is secreted by the pancreas into the small intestine. Another suitable suggestion is that insulin is too large to be absorbed. However, it would be digested before it reached the small intestine where most absorption occurs in the gut.

> **(c) (i)** Stem cells are undifferentiated cells that retain the ability to divide by mitosis. The cells that they produce can express certain genes to differentiate to become specialised cells. Some stem cells can give rise to many different types of specialised cell (embryonic stem cells); others can only give rise to a few types (adult or tissue stem cells).

ⓔ 3/3 marks awarded This is a thorough account of stem cells, including two of the different categories.

> **(c) (ii)** It is not necessary to use a large number of the cells as they should divide when put into the body. Also they can replace the β cells that have been lost and can continue to replace any that die in future. If the implanted cells survive, the person does not need to inject insulin.

ⓔ 3/3 marks awarded The information about stem cells in (c)(i) is used here to explain the advantage of using them in this treatment.

> **(c) (iii)** The stem cells are genetically identical to all other cells in the body so the immune system will recognise them as 'self' and cells such as killer T lymphocytes and phagocytes will not destroy them.

ⓔ 3/3 marks awarded This treatment involves using the person's own stem cells, rather than obtaining them from someone else, as happens with bone marrow transplants for leukaemia. This means there will be no tissue rejection. The idea that the cells will be genetically identical uses knowledge from the section on mitosis. This question also relies on knowledge of immunity from Module 4.

Question 7

(a) The diagram shows a molecule of lactose, which is a disaccharide.

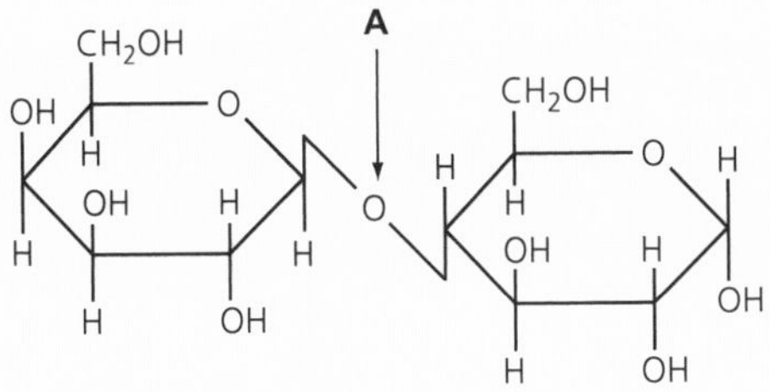

(i) Name the bond indicated by A in the diagram. (1 mark)

ⓔ When asked to name, you only have to give a very brief answer. In this case one word is enough.

(ii) Draw an annotated diagram to show how the bond is broken. (3 marks)

ⓔ Annotated means you should add notes, not just draw a diagram.

The enzyme β-galactosidase catalyses the breakdown of lactose. The chemical ONPG is colourless. It is also broken down by the same enzyme:

$$\text{ONPG + water} \xrightarrow{\text{β-galactosidase}} \text{ONP + galactose}$$

The product ONP is yellow, which means that qualitative and quantitative results can be taken to record the progress of the reaction shown above.

Some students investigated the reaction in which ONPG is broken down. They carried out a preliminary experiment using three test tubes, A, B and C. They put ONPG and water into the tubes as shown in Table 1. The tubes were placed in a water bath at 25°C and then a solution of β-galactosidase was added. The colours of the reaction mixtures were recorded after 10 minutes, with the results shown in Table 1.

Table 1

Tube	Contents of the tubes			Colour after 10 minutes
	Dilute ONPG solution/cm³	Water/cm³	β-galactosidase solution/cm³	
A	1.00	0.00	0.05	Yellow
B	1.00	0.05	0.00	Colourless
C	0.00	1.00	0.05	Colourless

(b) **(i)** Explain the results for tubes A, B and C. (3 marks)

(ii) Explain why it was necessary to include tubes B and C. (2 marks)

The students next investigated the effect of glucose and galactose on the activity of β-galactosidase. The solutions of the two sugars were of the same concentration. Five more tubes, D to H, were set up as shown in Table 2, which also shows the results.

Table 2

Tube	Contents of the tubes				Colour after 10 minutes
	2 cm³ sugar solution	Dilute ONPG solution/cm³	Water/cm³	β-galactosidase solution/cm³	
D	No sugar	1.00	0.00	0.05	Yellow
E	Galactose	1.00	0.00	0.05	Very pale yellow
F	Galactose	1.00	0.05	0.00	Colourless
G	Glucose	1.00	0.00	0.05	Yellow
H	Glucose	1.00	0.05	0.00	Colourless

(c) Explain why tubes D, F and H were included. (3 marks)

(d) Explain the results in tubes E and G. (2 marks)

ⓔ Parts (c) and (d) show how important it is to study the information provided very carefully and think about it before writing. Also consider your response to the command word. In these questions beware giving descriptions when explanations are required.

The students investigated the effect of changing the concentration of ONPG on the action of galactose on β-galactosidase. They set up seven identical test tubes each containing 1 cm³ of the ONPG solution with 1 cm³ of different concentrations of a galactose solution. These tubes were equilibrated at 25°C. To each tube in turn they added 0.05 cm³ of β-galactosidase and placed it into a colorimeter. They took an absorbance reading after 2 minutes for each test tube. They then repeated the procedure but without the galactose solution.

The students plotted their results to give the graph below.

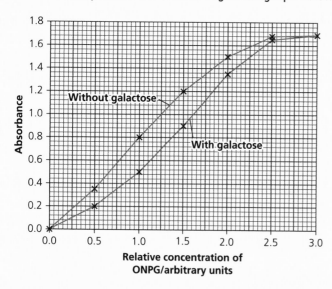

(e) Use the graph to describe the effect of galactose on the rate of the hydrolysis of ONPG by β-galactosidase. (3 marks)

(f) Explain the effects of galactose on the reaction. (3 marks)

Student answers

(a) (i) glycosidic

ℯ **1/1 mark awarded** This shows the importance of precision — some students are tempted to write 'covalent' as the answer. All of the bonds between the compounds that form macromolecules are covalent, so this is not specific enough.

(ii)

Glycosidic bond

+ H_2O

The glycosidic bond is broken by the addition of water — the circles show the H and OH from water after the bond is broken

ℯ **3/3 marks awarded** Note that the glycosidic bond is drawn slightly differently from the way it is shown in the question. Both are acceptable ways to show the bond in lactose.

(b) (i) A: The enzyme has catalysed the hydrolysis of ONPG to ONP (yellow).
B: There was no enzyme, so no reaction occurred — even a spontaneous one — so no yellow colour appeared.
C: As there was no ONPG (substrate) there could be no reaction.

ℯ **3/3 marks awarded**

(ii) These are controls. B and C make sure that the reaction is the source of the yellow colour, not one of the substances put into the reaction. For example, ONPG could just break down on its own when in solution.

@ **2/2 marks awarded** There is unlikely to be a mark for 'controls' on its own without any further explanation. Certainly at this level there is never a mark for using the phrase 'fair test' without any further explanation.

> **(c)** Tube D should give a colour that is the same as tube A. This can be used as a reference to compare with the other tubes (E to H). F and H are controls to make sure that the sugars do not react with ONPG, as that would invalidate the results from tubes E and G.

@ **3/3 marks awarded**

> **(d)** It appears that galactose inhibits the action of the enzyme, but glucose has no effect at all as the colour is the same after 2 minutes in tubes D and A.

@ **2/2 marks awarded**

> **(e)** The graph shows that at concentrations of ONPG up to 2.5 au the rate of reaction is less when galactose is added; for example at 1.5 au the absorbance with galactose is 0.90, but without it is 1.50. At the highest concentration of ONPG the rate of reaction is the same without galactose.

@ **3/3 marks awarded** Some comparative data from the graph are used to illustrate the description. In this case there is no need to include any units for absorbance. The key point about this graph is that it shows that the inhibitor is a competitive inhibitor because the effect is reversed by an increase in substrate concentration. Remember that the x-axis is for different reaction mixtures with different concentrations of ONPG; it is not time, as many can often imply by writing poorly phrased answers.

> **(f)** Galactose is a competitive inhibitor of the enzyme; it has a similar shape to ONPG and fits into the active site of β-galactosidase and prevents the entry of substrate molecules. But the inhibition effect decreases as the concentration of the substrate (ONPG) increases.

@ **3/3 marks awarded**

Question 8

Riboflavin is synthesised by plants, yeast and other microbes by a pathway of seven enzyme-catalysed reactions. This pathway is not present in mammals, which obtain riboflavin in their diet.

In all organisms, riboflavin is used to make a variety of coenzymes, including flavin adenine dinucleotide (FAD), which is the prosthetic group of the mitochondrial enzyme succinate dehydrogenase. This enzyme catalyses a dehydrogenation reaction during the Krebs cycle, which is a pathway in aerobic respiration.

(a) Explain the role of the coenzyme FAD in respiration. (3 marks)

ⓔ See Table 14 on p. 46 to help with answering this question.

(b) (i) What is meant by a prosthetic group? (1 mark)

(ii) Name another example of a prosthetic group. (1 mark)

ⓔ This question tests your knowledge of conjugated proteins.

(c) Riboflavin is a water-soluble molecule. Describe how it is absorbed by mammalian cells. (2 marks)

ⓔ This question tests another topic in this book. Be prepared to move from topic to topic within an exam question.

(d) In a student project, thin-layer chromatography (TLC) was used to carry out a qualitative analysis of the water-soluble vitamins in different brands of polyvitamin tablets.
Describe the steps that the student should take to identify the vitamins in each type of tablet. (8 marks)

ⓔ This question tests the skill of planning practical investigations by giving the steps to follow for a specific technique. During your course you should write many of these even if you never carry them out. In the case of chromatography it is a good idea to do some paper or thin-layer chromatography to learn the principles and the procedure and then be able to plan something like this.

Student answers

(a) Coenzymes take part in enzyme-catalysed reactions but they are not catalysts and they are not the substrate of the reaction. In this reaction FAD is changed (reduced) by the transfer of hydrogen ions from a substrate molecule. This is why the enzyme is called a dehydrogenase.

ⓔ **3/3 marks awarded** This question requires knowledge of Module 5, but there is enough in this guide to help you phrase an answer like this. Some students write answers to questions like these using only the information provided. This often happens when students do not know the particular part of the specification being tested.

(b) (i) A prosthetic group is a compound, not made from amino acids, that is bound to a protein to make a conjugated protein.

ⓔ **1/1 mark awarded**

(ii) haem (in haemoglobin)

Questions & Answers

ⓔ 1/1 mark awarded The question did not ask for the name of the conjugated protein, but it is probably a good idea to include that information (and it also reminds you). Zinc and carbonic anhydrase would be another possible answer. There are several places in the specification where bits of information do not seem to link with anything else. However, this is rarely the case as the Links sections in this guide point out.

(c) Water-soluble compounds are absorbed into cells by facilitated diffusion through carrier proteins or channel proteins as they cannot pass through the phospholipid bilayer. They could also be absorbed by active transport through carrier proteins.

ⓔ 2/2 marks awarded Both answers could be right. We do not have enough information to know whether riboflavin is absorbed by either of these two methods. If, as seems likely, riboflavin is converted into something else as soon as it enters, facilitated diffusion may be most likely as there will always be a concentration gradient across the membrane into the cells. Thus cells do not need to use valuable energy to absorb this compound by active transport.

(d) Here are instructions for carrying out thin-layer chromatography on the vitamin tablets.

1   Dissolve each of the tablets in water.

2   Rule a line with a pencil across the TLC plate (10 mm from the base and at least 5 mm from the edge). Mark a spot 10 mm from an edge for the solution made from the first tablet. Put other spots at 10 mm apart across the line for the other tablets.

3   Take a fine pipette and put it into the liquid and then touch the spot. Let the spot dry and keep doing this until there is a dark spot that does not spread more than 3 mm across.

4   Stand the TLC plate in a jar containing a solvent to a depth of less than 10 mm.

5   Cover the top of the jar and leave until the solvent front reaches to within 10 mm of the top.

6   Take out the TLC plate, mark the solvent front with a pencil and dry it.

7   Look at the plate under UV light. The different vitamins appear different colours. Draw around the coloured areas.

8   Measure the distance from the start line to the solvent front. Then measure the distances from the start line to the middle of each spot and use these to calculate the retention factor values (Rf):

$$Rf = \frac{\text{distance travelled by each spot}}{\text{distance travelled by solvent front}}$$

9   Compare the colours and the Rf values with those published to identify the vitamins.

ⓔ **8/8 marks awarded** Wow! This answer shows a very good understanding of one of the practical techniques specified in Module 2. Writing a series of numbered steps is a good approach to a planning question like this. Do not use bullet points, but use numbers instead so they can be referenced, as in 'repeat step 3'. This is much easier to write (and to mark) than dense prose.

ⓔ Succinate dehydrogenase is a very synoptic enzyme:

■ It is situated in a membrane — the inner mitochondrial membrane — so the reaction it catalyses occurs on the surface of a membrane.
■ It catalyses a reaction of the Krebs cycle — one of the pathways of respiration that occurs in mitochondria.
■ It has a prosthetic group, which is also a coenzyme.
■ The compounds malonic acid and oxaloacetate are competitive inhibitors.
■ It catalyses a reaction that is not a hydrolysis reaction — it involves the transfer of hydrogen ions from a substrate molecule to FAD.

Question 9

The enzyme glucose isomerase catalyses the interconversion of glucose and fructose:

$$\text{glucose} \xrightleftharpoons[]{\substack{\text{glucose} \\ \text{isomerase}}} \text{fructose}$$

This enzyme is immobilised onto the surface of small resin beads. A student investigated the activity of this immobilised enzyme by placing a known mass of beads into a test tube containing $10\,cm^3$ fructose solution and placing this into a thermostatically controlled water bath at 65°C. The student took small samples of the reaction mixture at intervals and tested them with reagent strips that detect glucose.

The table shows the student's results.

Time/s	Colour of reagent strip	Concentration of glucose/ mg $100\,cm^{-3}$
0	Blue-green	0
60	Pale green	100
120	Green	250
180	Dark green	250–500
240	Brown	1000

(a) Discuss the advantages of using reagent strips rather than Benedict's solution in this investigation. *(4 marks)*

ⓔ Read the introduction very carefully before reading the part questions. This is a very good example of why it is necessary.

(b) State **two** limitations of the method and results described above that may have affected the quality of the results obtained.

For each limitation:
■ explain how it influenced the quality of the results
■ describe how you would modify the procedure to overcome the limitation *(6 marks)*

ⓔ This tests your skills of evaluating. Read what you have been told again and plan the whole answer — limitation, explanation and modification — before writing.

Student answers

(a) Using reagent strips is much easier than taking a sample, boiling it with Benedict's solution and waiting to see the result, especially if five samples are going to be taken. Reagent strips give quantitative results, whereas it is quite difficult to do that with Benedict's solution. Also, Benedict's solution reacts with all reducing sugars to give a positive result, so it would be impossible to tell that glucose was being produced.

ⓔ **4/4 marks awarded** This student has spotted that Benedict's solution cannot be used in this investigation as it is not a specific test for glucose, whereas the test strip for glucose has an enzyme that accepts glucose as a substrate but not fructose (a good example of specificity).

(b) 1 The student carried out the investigation only once, so does not know how much confidence to put in the results. I would repeat the experiment at least twice in exactly the same way as before, for example using a water bath at 65°C and taking samples at exactly the same intervals. The results could be compared to see if they are the same and are therefore repeatable.

2 It is difficult to match the colours of the pads on the strips to the colour chart. Also the student thought that at 3 minutes the colour was between two of those on the chart and therefore gave the concentration as a range. This is no good for plotting graphs. I would use a biosensor to give precise readings within the range of accuracy of the device, for example $110\,\mathrm{mg\,cm^{-3}} \pm 5\,\mathrm{mg\,cm^{-3}}$.

ⓔ **6/6 marks awarded** You should be prepared to evaluate experimental procedures and data; also to assess whether the evidence from an experiment supports a hypothesis or not. In this case, you are asked to evaluate a method that you may not have carried out. However, you should have used test strips and know that results are taken by matching the colour on the strip with a colour chart.

Answers to multiple-choice questions

Question	AS	A2
1	C	D
2	B	A
3	B	C
4	C	D
5	C	C

Knowledge check answers

1 The formula is:

magnification = $\dfrac{\text{width of image}}{\text{actual width}}$

$75 \times 1000 = 75000\,\mu m$

$\dfrac{75000}{40} = 1875$

magnification = ×1875

(Remember to show your working in questions like these.)

2 The formula is:

actual length = length of image/magnification

$14 \times 1000 = 14000\,\mu m$

$\dfrac{14000}{2000} = 7\,\mu m$

(When calculating magnifications and actual sizes always convert measurements into micrometres (μm) first.)

3 The areas stained blue-black contain starch; the areas stained yellow do not. The cells must contain starch grains. Cell walls, cytoplasm and the nucleus stain yellow.

4 Cell in Figure 3 is approximately 27 μm. Cell in Figure 4 is approximately 140 μm. Actual widths depend on where you have measured the two cells.

5/6 You could incorporate both tables, as here. Always remember to include a column for the features you are comparing.

Feature	Prokaryotic cell	Eukaryotic cells Plant cells	Eukaryotic cells Animal cells
Typical size/ μm	0.5–3.0	40–60	≈ 20
Capsule	Found in some	✗	✗
Cell wall	✓ (murein, not cellulose)	✓ (made of cellulose)	✗
Membrane-bound organelles	✗	✓	✓
Vacuoles	✗	Large central vacuole	Small (vesicles)
Ribosomes	Smaller: 20 nm/70 S	30 nm/80 S	30 nm/80 S
Nucleus	✗	✓	✓
DNA	Bacterial chromosome is a ring of DNA in the cytoplasm	Chromosomes made of linear DNA found within the nuclear envelope	

There are other differences you could include (see Table 3 on p. 14).

7 Solvent in cytoplasm and body fluids, such as blood and tissue fluid; reactant in hydrolysis reactions (e.g. in digestion); coolant in sweating and panting.

8 In α-glucose the –H is above the –OH on carbon 1. In β-glucose it is the reverse.

9 α-glucose has six carbon atoms; ribose has five. On C1 the –OH is below the –H.

10 maltose, glycogen, amylopectin, amylose — all α-glucose; sucrose — fructose and α-glucose; cellulose — β-glucose

11 Amylose is a polymer of α-glucose; cellulose is a polymer of β-glucose. Amylose has a helical shape and is compact, so is suitable for storage. Cellulose is a straight-chain molecule with its monomers arranged alternately at 180° to each other. It forms many hydrogen bonds with surrounding cellulose molecules to form tough microfibrils in cell walls for support.

12 Not a polymer; not branched; contains at least two different sub-unit molecules; these are glycerol and fatty acid; glycogen is made only of α-glucose; triglyceride contains ester bonds (not glycosidic bonds).

13 a Unsaturated fatty acids have one or more double bonds in the hydrocarbon chain.

b Hydrophobic and hydrophilic 'ends'; as part is hydrophobic cholesterol is not water soluble; made of carbon rings (whereas phospholipids and triglycerides are not).

14

Features	Triglycerides	Phospholipids
Contains phosphate group	✗	✓
Charged/polar	✗	✓
Number of fatty acids	3	2
Functions	Energy storage, buoyancy, protection, thermal and electrical insulation	Formation of bilayers in cell membranes

15 There are 20 different monomers (amino acids) for making proteins. There are very few for making polysaccharides (in this course you know of two — α-glucose and β-glucose).

16 Nucleotides of a new polynucleotide are assembled against the template strand by base pairing: A–T, T–A, C–G and G–C. So if the template strand is ATCGTTA, the newly synthesised strand will be TAGCAAT.

17 a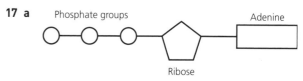

b ATP + H₂O → ADP + Pi

18 Histidine (His): on the coding strands, CAT and CAC; on the template strand, GTA and GTG. Phenylalanine (Phe): on the coding strands, TTT and TTC; on the template strand: AAA and AAG.

You may find other terms used for the polynucleotides (or strands) of DNA:

- coding — sense
- template — antisense, reference

19 A two-base code would only code for $4^2 = 16$ amino acids.

20 met–leu–ala–ile–ala

21 Reading errors may give rise to a different sequence of amino acids — especially if they are in the first base of a codon.

22 DNA has two antiparallel strands. 3′ and 5′ indicate the different ends of a polynucleotide. They refer to the carbon atoms on the sugar, deoxyribose.

23 Universal — the genetic code is the same in all organisms and in viruses. Degenerate — most amino acids are coded by more than one triplet. Tryptophan and methionine (Met) are the two exceptions. Non-overlapping — each base is part of a single triplet; the ribosomes 'read' the mRNA sequence in groups of three bases that are separate from each other in the 'reading frame'.

24 'Fluid' refers to the phospholipid bilayer and 'mosaic' refers to the proteins that are situated within the bilayer. Phospholipids and proteins are free to diffuse within the membrane, although phospholipids tend not to 'flip' between one monolayer and the other. The two faces of the transmembrane proteins face different ways — into the cytoplasm and into the outside environment — and they tend not to 'flip' over either.

25 Cells secrete substances by exocytosis. These substances interact with receptors on the surfaces of adjacent cells (e.g. histamine and chemicals released at synapses) or cells at a distance (e.g. hormones).

26 Molecules and ions diffuse through channel proteins or carrier proteins in facilitated diffusion. In simple diffusion molecules pass through the phospholipid bilayer.

27 The movement of the cilia requires much energy; mitochondria produce ATP that provides energy to the cilia.

28 In mitosis the number of chromosomes remains the same in the daughter cells as in the parent cell; in meiosis the number halves. There is one division of the nucleus in mitosis; there are two in meiosis; homologous chromosomes pair together in meiosis; they do not pair in mitosis. There is crossing over and independent assortment in meiosis but not in mitosis. The nuclei of the daughter cells are genetically identical in mitosis; there is variation among the nuclei produced in meiosis. Mitosis is involved in growth, asexual reproduction, repair and replacement of tissues. Meiosis occurs during the formation of gametes (sex cells) in animals and the formation of spores in plants.

29 As a result of meiosis I, each chromosome has two possibilities. Therefore the total number of different combinations of the chromosomes is $2^{23} = 8388608$.

Note: **Bold** page numbers indicate defined terms

Index

light microscopes (LMs)   6, 7
light microscopy   8
lipids   20–23
lock-and-key model   40
lysosomes   11, 12
lysozyme   26

M

macromolecules   **16**
magnification   **7**
maltose   18, 19
mammals
    cell specialisation   59–60
    cells and tissues   60, 61
    mitosis   57–59
    organs and organ systems   62
meiosis   **63–67**
membranes
    fluid mosaic structure   49–52
    movement across   52–54
meristems   **56**
messenger RNA (mRNA)   32
    transcription   35–36
    translation   37–38
metabolism   **40**
metaphase   57
metaphase I and II   64, 65, 66
microscopy   6–10
mitochondria   11, 12
mitosis   **56–59**
molecules   15–29
monomers   **16**, 19, 20
monosaccharides   17
mosaic   51
muscle   61

N

neutrophils   59, 60
nitrogenous bases   30–31, 32
non-competitive inhibitors   47, 48
non-reducing sugar test   28
non-reversible inhibitors   46, 48
non-sister chromatids   **63**, 66
nuclear envelope   11, 12, 58
nucleic acids   30
nucleotides   30
    'energised'   34

O

organelles   10–13
organs and organ systems   62
osmosis   **53**

P

passive transport   52
pentose sugars   17, 30
pH and enzyme activity   44
phloem   61, 62
phosphate group   22, 34
phosphodiester bond   34
phospholipids   22, 50, 51
plant cells   7, 10, 11, 60
plant tissues   61–62
    differential staining   8
polymer   **16**
polynucleotides   31
    DNA replication   32–33
    transcription   36
polysaccharides   17, 20
primary structure   24, 25
prokaryotic cells   13–14
prophase   57, 58
prophase I and II   64, 65, 66
prosthetic groups   26
proteins   23–24
    amino acids   23–25
    fibrous   26–27
    globular   26
    membranes   50
    structure of   24–25
protein synthesis   35–38
protein test   28
purine bases   30, 31
pyrimidine bases   30, 31

Q

quaternary structure   24, 25, 38

R

random assortment   63, 66
reducing sugar test   28
replication of DNA   32–35
resolution   **6**, 7
reversible inhibitors   46–48
ribose   17, 32
ribosomal RNA (rRNA)   32
ribosomes   11, 12, 14, 32
    role in translation   37–38
RNA (ribonucleic acid)   30–32, 34
    protein synthesis   35–38
root hair cells   60

S

scanning electron microscope (SEM)   7
secondary structure   24, 25, 26

sieve tube elements   62
simple diffusion   53
sister chromatids   **57–58**
specialisation   59–61
sperm cells   60, 61, 64
squamous epithelia   60, 61
starch test   28
stem cells   **56**
substrate concentration   44–45
substrate molecule, enzymes   40–41,
    46–47
sucrose   18, 20
sugars   17, 19, 28

T

telophase   57, 58
telophase I and II   64, 65
temperature and enzyme activity
    42–43
tertiary structure   24, 25, 28
    enzymes   40, 42
tests, biological molecules   28
thymine   30, 31, 32
tissues   60–62
transcription   **35–36**
transfer RNA (tRNA)   32
    amino acid activation   36–37
    translation   37–38
translation   **35**, 37–38
transmission electron microscope
    (TEM)   7, 9–10
triglycerides   20, 21–22

U

units, microscopy   6
universal genetic code   35
urease   46, 48

V

vessel elements   61, 62

W

water molecule   15–16
water potential   **53**

X

xylem   61, 62

Y

yeast   59